Unravelling animal behaviour

D0715289

Unravelling animal behaviour

MARIAN STAMP DAWKINS

*Mary Snow Fellow in Biological Sciences,
Somerville College, Oxford*

LONGMAN

Longman Group Limited
Longman House, Burnt Mill, Harlow
Essex CM20 2JE, England
Associated companies throughout the world

First published 1986

British Library Cataloguing in Publication Data
Dawkins, Marian Stamp
Unravelling animal behaviour.
1. Animal behaviour
I. Title
591.51   QL751
ISBN 0–582–44691–0

Library of Congress Cataloguing in Publication Data
Dawkins, Marian Stamp.
Unravelling animal behaviour.

Bibliography: p.
Includes index.
1. Animal behaviour.   I. Title.
QL751.D375 1986   591.51   85–10230
ISBN 0-582-44691-0

Printed and bound in Great Britain at
The Bath Press, Avon

To Max Stamp
(1915–1984)

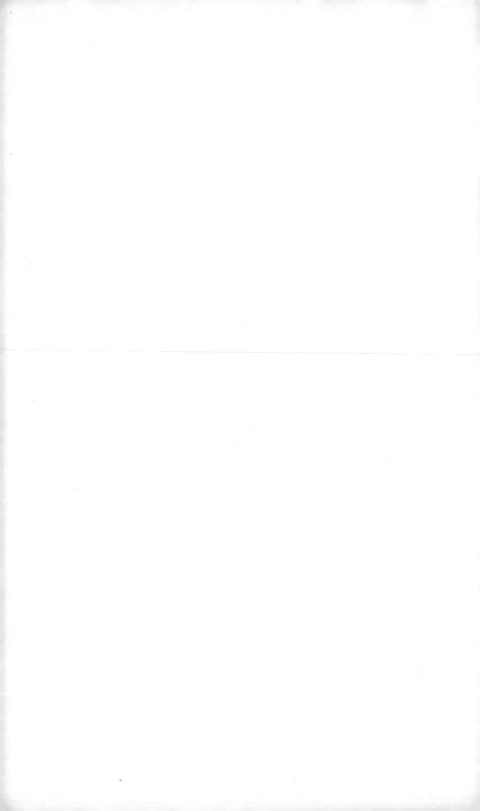

Contents

Preface ix

Acknowledgements x

Introduction 1

1 Adaptation *3*

2 Optimality *20*

3 Inclusive Fitness *33*

4 Genes and Behaviour *45*

5 Innate Behaviour *55*

6 Some Obstinate Remnants: Instinct, Displacement Activities, and Fixed Action Patterns *67*

7 The Machinery of Behaviour *83*

8 Communication *100*

9 Evolutionarily Stable Strategies *112*

10 Sexual Selection *131*

Epilogue 147

References 149

Preface

I have written this book as a guide, a companion to help unravel some of the difficulties and confusions which sometimes occur in what people have thought and written about animal behaviour. Most of the difficulties, if it is any comfort, have trapped me at some time or another and so I write as one who has stumbled over but eventually found a way out of difficulties rather than as a superior being who has never encountered any. Please do not look to this book for a review of the whole subject or even parts of it as it is not, and does not pretend to be, a comprehensive account of all animal behaviour. It is to be read alongside the numerous textbooks that now exist on the subject. It is like the mortar between the bricks of a wall. It is supposed to bind and make sense of an existing body of knowledge, not to provide all that knowledge itself.

The book is primarily intended for students taking a course in animal behaviour, but I hope it can be read by anyone with an interest in the behaviour of animals whether or not they have studied the subject before. References are given in the usual way at the end of the book and a separate list of key articles is given at the end of each chapter, so that anyone who wishes to gain a fuller and more comprehensive picture of the work that has been done can do so. But to make it easier to read, the number of references in the text itself has been deliberately kept down.

My own understanding of animal behaviour has been greatly helped by conversations with a large number of people, but particularly by those with Alan Grafen, who has the most annoying habit of always being right. Donald Broom, Michael

Hansell and Felicity Huntingford were outstandingly conscientious referees and Michael Rodgers of Longman provided the right encouragement at the right times.

Marian Dawkins
Somerville College, Oxford
February 1985

Acknowledgements

We are grateful to the following for permission to reproduce copyright material:

Cambridge University Press for our fig 3 from fig 1 p 256 (Lorenz 1950); Chapman and Hall for our fig 5 from fig 3.2 p 38 (Sales and Pye 1974); Alan R. Liss, Inc. (New York) for our fig 6 from p 318 (Simmons and Vernon 1971); the author, Professor D. Weihs, and Plenum Publishing Corporation (New York) for our fig 1 from fig 1 (Weihs 1975).

Introduction

Animal behaviour is all around us. There are animals in our houses, on farms and in the natural world outside. Television shows us the lives of animals we might otherwise never see. The study of animal behaviour – sometimes called ethology or behavioural ecology or sociobiology (when the emphasis is on social behaviour) – is a popular and flourishing subject. It has an immediate fascination to a wide range of people as well as practical applications in agriculture and animal welfare. And yet, despite the popularity of the subject and the apparent ease with which animal behaviour can be studied, there are difficulties, stumbling blocks that get in the way of a proper understanding.

None of these difficulties is insuperable. Any of them can be mastered by anyone prepared to give them a little thought and this book is a guide to doing just that. Sometimes we will find clear-cut answers, at other times the answer will depend on the definition that is being used and at yet other times we come up against the limits of anyone's knowledge. The chapters are arranged so that we both begin and end with problems to do with the evolution of behaviour.

The meaning of 'adaptation' and the difficulties with substantiating adaptive hypotheses about behaviour start us off in Chapter 1. Chapter 2 continues the discussion of natural selection by exploring the question of whether 'adaptation' ever leads to 'optimality'. 'Inclusive fitness', which is the subject of Chapter 3, provides a clear-cut example of a widespread misunderstanding which can be easily cleared up with a little thought. The genetic basis of behaviour, having been assumed during the first three chapters, gets an airing in its own right in Chapter 4, where the problematical relationship between genes and behaviour is brought out into the open.

'Innate' behaviour, and what that might mean, follows naturally from the discussion of genes in Chapter 5. But then, having pulled up 'innate' behaviour for examination, we find that it has brought with it its old alliance with 'instinct' and some classical ideas about displacement activities and fixed action patterns. If these were long dead ideas of merely historical interest, we could have glossed over them and said no more. But because some of them still persist and still muddle peoples' thinking to the present day, Chapter 6 is spent examining them. Chapter 7 turns to some more recent, but still controversial, ideas about the 'machinery of behaviour' and whether physiology or a 'whole animal' approach is the best way to study it.

Next, in Chapter 8, we look at communication and signalling between animals, but here we find such problems with the way people have used words that we have to spend almost the whole chapter sorting out definitions. We return to the effect of natural selection on behaviour in Chapter 9 by looking at evolutionarily stable strategies and the benefits and problems they have brought. Finally, in Chapter 10, we focus on some of the most persistent evolutionary problems of all – those to do with sex and sexual selection. After all our concern with problems of definition and misunderstanding, we thus end with genuine mysteries.

1
Adaptation

Pink-footed goose (*Anser fabalis brachyrhynchos*).
Photograph by Tony Allen.

It may seem odd to start a book called *Unravelling Animal Behaviour* not with animal behaviour but with the theory of natural selection. The reason for doing it is not just that many of the questions people ask about animal behaviour concern the way behaviour has evolved and the way natural selection has acted on it. It is because more confusion and misunderstanding have arisen over the natural selection of behaviour than almost anything else. This may come as a surprise to anyone who thought the theory of natural selection was simple and straightforward. But misunderstandings and misapplications of it are to be found everywhere. People have misunderstood the idea of 'adaptation' and some of them have quite wrongly assumed that theories of adaptation are untestable and nothing more than 'Just-So' stories (Gould and Lewontin 1979). There have been misunderstandings about what it means to say that animals behave optimally (Ch. 2) and, as for 'inclusive fitness' (Ch. 3), almost all current textbooks are misleading. So it is not really all that strange to begin our study of animal behaviour with a long hard look at the theory of natural selection itself.

Natural selection and the problem of alternatives

The theory of natural selection as put forward by Charles Darwin is sometimes seen as an assertion that all animals are adapted to their environments, with characteristics which 'fit' them to their way of life. On this view, giraffes are said to have long necks to reach the tops of trees and stick insects to look like twigs to avoid being eaten by birds.

But this is not, in fact, what Darwin's theory of natural selection is all about. His theory is not about animals simply surviving and reproducing. He did not suggest that animals and plants were engaged in a private fight with their physical environments, important though it may be for them to 'battle' with heat or cold. Darwin's theory is, by contrast, about organisms surviving *better than* their competitors. He saw animals engaged in a struggle to exist and reproduce in which the best plant or animal won. It was not enough to be good at surviving. The important thing was to be better than the competition.

So, apparently simple questions such as 'Why does a giraffe have a long neck?' or 'Why do rabbits have such big ears?' are much more difficult to answer than they appear at first sight. If large ears in rabbits evolved because large-eared rabbits did better in competition with rabbits with other sorts of ears, what were these other rabbits like? And why did they not do as well? Did they have smaller ears which could not pick up distant sounds as well or did they have larger ears which got caught up in the sides of burrows? Did they have rudimentary ears or ones of a different shape altogether?

Putting the problem in this way enables us to see that any question about adaptation really involves at least two separate questions. Firstly, there is the question of what alternatives were available for natural selection to choose between. Secondly, there is the question of why one alternative did better than the others. In many cases, the competitors which failed in the struggle for existence will no longer be around; but if we accept Darwin's theory, we have to postulate that, at least sometime in the past, they did once exist and that there was a reason for their demise.

The trouble with this is that we can postulate alternative kinds of animal indefinitely – pigs with wings, green rabbits, and so on. But whether these actually were the raw material of evolution is extremely difficult to determine. In other words, it is very hard to know whether they did once exist but were found wanting in the struggle for life or whether such mutations never have and never could have arisen within that species.

If they never have existed, as rabbits with little metal wheels have certainly never existed, then they are not one of the alternatives available for selection. But whereas we can confidently rule out little metal wheels as part of the rabbit alternatives for evolution and equally confidently include rabbits with slightly larger ears than average (because they exist), there is a grey area in between. Could mutations for rabbits with differently shaped ears, or three ears arise? Indeed, have they? Were these mutants once alive and were they selected against? Or have the mutations never arisen, never given the rabbits possessing them a chance to prove themselves, one way or the other, in competition with other rabbits?

The second question about adaptation can be asked only when we have somehow overcome these problems and decided what alternatives were available for selection. The question is this: *why* did one alternative do better than the others? What happened to the losers? To put it at its most crude, because of what have the failures in the struggle for existence died or failed to reproduce?

Notice that this is not a question about *whether* natural selection has taken place. It is about the fates of winners and losers. Given that there were rabbits with ears of many different shapes and sizes, why did the big-eared ones have the edge over smaller-eared ones? Was it perhaps because they could hear better and so were less likely to die from predation? Or was it because they could regulate their temperature better and so were less likely to die of heatstroke?

To say that something is an 'adaptation' in an animal alive now implies that we can distinguish between the various hypotheses about causes of death and reproductive failure in rivals that may have existed many millions of years ago. It is like a murder mystery that took place a long time in the past, so long ago that nobody can remember who lived and who died or whether they died by the sword or by being poisoned. The only remaining clues, apart from a few fossilized bodies, are the present-day living descendents of those who managed to survive some unspecified disasters of the past.

It is from such unpromising material that we have to derive our hypotheses about adaptation and to work out what befell those that were unsuccessful. Contrary to what is sometimes believed, however, this does not mean that all we can do is to speculate with no possibility of obtaining any sort of evidence. There are in fact four well-established methods for putting the study of adaptation on a firm basis and we will now look at these. Two of them could be described as 'direct' or 'reconstruction' methods, in that they involve trying to observe death and reproductive failure in present-day animals as indications of what probably happened in the past. We either compare two existing forms of the same species (discussed under 1 below) and see which one fares better or we compare a single variety with some artificially contrived, man-made alternative (see 2 below).

The other two methods, which are 'indirect', spare us the need to actually observe killing and death and concentrate instead on the results of natural selection. If present-day animals are the result of a struggle for existence that took place in the past, it may be possible to make inferences about what was going on then by studying the descendents of those victors now. Natural selection in the past is studied through its legacy in present-day animals either through cross-species comparisons (see 3 below) or through uncovering 'design features' of structure or behaviour (see 4 below).

Evidence for adaptation

1. Making use of existing genetic variation

The most straightforward of the direct methods is to exploit the fact that, even in present-day species, not all individuals will be exactly alike. There will be variation in size, colour, and perhaps various aspects of their behaviour that will lead us to ask whether any of these characteristics enable their possessors to survive and reproduce better. The classic example of this is the colour variation in the peppered moth (*Biston betularia*). The natural occurrence of light and dark forms of this moth enabled Kettlewell (1955) to show that in industrial areas, where a sooty atmosphere meant that most of the tree-trunks were dark and not covered with lichen, the dark form was better protected against bird predators than the lighter form. The light form was, in turn, better camouflaged than the dark moths where lichen tended to cover the tree-trunks. In this case, the colour adaptation was studied through a direct study of the death of the uncamouflaged moths. Moths of different colours were experimentally released in different areas and birds were then seen to fly onto the trunks and eat insects that stood out from the background, leaving behind the ones whose colour enabled them to resemble just another patch of tree. Death took place under the eyes of the observers.

There are a number of important features which we should notice about this familiar and often-quoted example. In the first place, it is known that the differences between the dark and the light forms have a genetic basis and so their survival and death can give a clear indication of the gene frequencies to be found in the next generation. Birds can almost be seen to remove some genes and leave others.

The same cannot be said, however, for many other cases where one kind of animal survives and another is killed, and where it would be quite easy to imagine that natural selection was being observed in action. In black-headed gull colonies, the birds nesting at the centre of the colony have a higher reproductive success than those nesting at the edge (Patterson 1965). Predators such as gulls, crows, and foxes find it much easier to take young and eggs from nests at the edge of the colony because the massed attacks of the parent gulls protect the centre nests. But dead chicks or adults at the edge of the colony do not necessarily indicate that natural selection is taking place because we do not know whether there are genetic differences between gulls nesting in different places. They could all be genetically similar as far as nest-site selection is concerned but nest in different places depending on what was available. In fact, the central, safer nest sites do tend to be occupied by older gulls, with the younger ones being pushed to the more dangerous outside. We cannot conclude from the fact that chicks in outside nests are more likely to be killed that one genetic form is being penalized and another favoured, as we can with the moths.

A second point to be noticed about the peppered moth example is the fact that the effects of natural selection could be 're-staged' artifically by importing alternative uncamouflaged moths in from another area where they would have been camouflaged. (The paler form also persists at a low frequency even in heavily polluted industrial areas.) But in many other cases, such convenient alternatives cannot be found at all. Frequently, only one variant will remain, the others having been eliminated by the very process – natural selection – that we want to study. This may foil our attempts to re-stage evolution, but it is hardly surprising. If natural selection has been acting in the way we think it has, unsuccessful alternatives should no longer be available for us to do our comparisons with.

So, direct comparisons of the peppered moth kind, although an ideal and very revealing way of studying adaptation, are in many cases not feasible. Two genetic alternatives may simply not be available for comparison and so a second, somewhat different method has to be adopted as a substitute.

2. Using artificially produced variation

Although it may not always be possible to find the required unsuccessful variants occurring naturally, it may nevertheless be possible to create them artificially. These artificial man-made 'mutants' can then be used to show how good real animals are by comparison. There is, of course, the obvious objection that the artificial variants may not precisely mimic the mutations that have, or could ever have, arisen naturally within a population, but with a little ingenuity and knowledge of the species concerned, very plausible 'mutants' can be devised.

Niko Tinbergen and his colleagues (1967), in a pioneering experiment along these lines, set out to discover why black-headed gulls remove empty eggshells from their nests. After a chick has hatched, a parent bird will pick up the shell in its beak and fly some distance from the nest with it before dropping it well away from the nest area.

Tinbergen wanted to know why the gulls do this, why, that is, gulls that remove eggshells are at an advantage compared with those that do not. This was particularly problematical because the newly hatched chicks are very vulnerable to being eaten by crows and herring gulls while the parent is in the act of taking the eggshell away. In addition, some black-headed gulls prey selectively on the wet chicks of their neighbours that have just come out of the egg. The possible explanations for the evidently risky behaviour of removing shells appeared to be that if the eggshells remained in the nest they might damage the chick, or that one might slip over an unhatched egg and prevent the next chick from coming out, that it might make it more difficult for the parent to brood, that shells might harbour disease, or that they might attract the attention of predators. Notice that – in line with what we were saying earlier about adaptation – each of these hypotheses is about a different possible *cause of death* in chicks if gulls did not remove eggshells. Discovering the nature of the adaptation means discovering which of them was the most likely to occur.

Because all black-headed gulls, as far as is known, do remove eggshells, there is no way of finding out what the advantage of the behaviour is using natural mutants that do not show this behaviour. So Tinbergen devised some artificial mutants – man-made gull nests complete with eggs from which no eggshells had been removed – and compared them with other nests with eggshells removed to different distances. He found that the least successful 'mutants' were those nests with eggshells left in them or near them. Crows and herring gulls flew down and took the unhatched eggs that remained. Nests with eggshells well away from them were much less easily spotted by the predators.

All this strongly suggests that if a real mutant black-headed gull that did not remove eggshells existed in the past, its reproductive success would have been lowered because eggs and young would be more likely to be eaten by predators. Eggshell removal is therefore most probably an anti-predator adaptation, death of offspring through predation being the most likely reason why past rivals lost out, as opposed to death of offspring through disease harboured in the shells or injury from the sharp edges.

Even this method, however, is not altogether satisfactory, partly because the experiments are very difficult to do in practice and partly because the artificial 'mutants' are only an approximation to what goes on or what has gone on in nature. If we happen to have thought up the wrong mutant, the answers to our comparison could be quite misleading. Tinbergen chose to compare removing eggshells with not removing eggshells using shells that were white inside (as real gull eggs are), but a similar comparison using shells that were green and camouflaged on the inside as well as the outside might have given very different results. It is plausible to suppose that the past gull rivals had eggs that were white inside like present-day gulls and were distinguished from their successful competitors only by a certain slovenliness over keeping their nests tidy – but that is a supposition. The original struggle for existence may have been between participants that were rather different from the 'removers' and 'non-removers' that were re-created in the present day.

As well as this direct evidence, however, there is another way of studying adaptation. This is altogether a gentler approach. The two methods we will now consider do not involve actually having to see predation or starvation taking place. They assume that the death and killing occurred in the past but they do not demand a reconstruction in the present. Their raw material is the animals of the present day, the victors of those long-ago events. By looking at what makes a victor, they attempt to work out the nature of the victor's advantage. Adaptation is studied not through reconstruction, but through its effects.

3. 'The Comparative Method'

Although all methods for studying adaptation are to some

extent comparative (success is always relative to something else), there is one that has come to be known as the 'Comparative Method'. 'Comparative' here means comparisons between species (or less commonly, comparisons between populations of the same species) which are living in different areas or niches and are therefore subject to different selection pressures. Now, looking at the effects of different selection pressures is precisely what we are interested in in our pursuit of adaptation, but it may not at first sight be clear how making comparisons between totally different species throws any light on it. The rationale goes something like this. Suppose that species A lives in an environment in which there are not very many predators, whereas species B, which is closely related to it, lives in another environment where there is constant danger from predation. Differences between the behaviour of species A and species B might tell us something about the behaviour that B has evolved specifically as an anti-predator adaptation. Of course, we have to be careful with such comparisons. There may be other differences between the environments of A and B besides the obvious one of predation pressure. But the more species we look at and the greater the number that behave like A when they suffer little risk of predation and like B when predation is heavy, the more confident we can be that we have correctly identified the crucial selective force.

E. Cullen (1957) made a detailed comparison between the behaviour of the kittiwake and various other closely related gull species. The importance of her study lies in the fact that there is a crucial difference between the environment of the kittiwake and that of these other species. Most gulls nest on the ground whereas kittiwakes nest on steep cliffs, out of reach of crows and herring gulls, which are dangerous predators to the young of ground-nesters. Sometimes even the kittiwakes themselves have to make several attempts before being able to land on their own nests, perched as they are on narrow ledges; but at least their young are virtually free from predation.

Kittiwake nests are extremely messy. The parents do not remove the empty eggshells and their nests are made highly conspicuous by their white droppings. This is very interesting confirmation that eggshell removal in ground-nesters is indeed an anti-predator adaptation, because all the other potential dangers such as the shells harbouring disease or injuring the chicks would seem to apply just as much in kittiwakes as in other gulls. Yet, released from the pressure of predation, eggshell removal is absent. The fact that this behaviour occurs in close relatives of the kittiwake strongly suggests that similar tendencies could arise (and perhaps have previously arisen) in the kittiwakes themselves. A kittiwake that removes eggshells, like a ground-nesting gull that does not, was a plausible candidate for selection. The non-occurrence of this behaviour in kittiwakes and its very striking occurrence in other species clearly points to the action of selection operating through predators (Cullen 1960).

Such comparisons become even more convincing when larger numbers of species are considered. Gannets, which, like kittiwakes nest in inaccessible places, are similar in not removing eggshells and having generally conspicuous nests. In fact, with respect to many of its nesting habits, the kittiwake is more like a gannet, to which it is only distantly related, than to other gulls which nest on the ground (Nelson 1967).

The Comparative Method, particularly when applied quantitatively across a wide range of species, is an extremely powerful way of showing up the action of natural selection. By searching for correlations between the environments in which animals live and the behaviour which they show, it is possible to see how their behaviour helps them to survive and reproduce. Otherwise puzzling aspects of what animals do – such as flying up with empty eggshells and dropping them – can often be understood by showing which species show the behaviour and which do not. The Comparative Method is not without its difficulties, however; the chief of these is that 'correlation' does not necessarily mean 'causation'. In other words, just because there may be a correlation between susceptibility to predators and removing eggshells, it does not follow that predation causes the difference between the behaviour of kittiwakes and that of other gulls. It could be some other factor, such as the temperature of their nest sites or some unsuspected disease risk peculiar to kittiwakes. We have to be careful not to jump to conclusions (Ridley 1983; Clutton-Brock and Harvey 1984).

Despite this, the distribution of a behavioural trait – that is, its presence in some animals pursuing one mode of life and its absence from other, perhaps closely related animals doing something else – is one of the most important kinds of evidence we can have in our pursuit of adaptation. What we see now are the survivors, the successful participants in selective processes that led to the demise of so many others. Those that are left behind carry the hallmarks of what those processes must have been. By comparing success stories in different environments, we gain, by implication, an idea of what the losers were like and why they failed.

4. Adaptation through 'design features'

There are enormous practical difficulties in relating past variations in morphology or behaviour to variations in the numbers of surviving offspring produced. Yet this is what we have to aim at if we want to study adaptation. Some of these practical difficulties concern, as we have already seen, the problem of recreating animals with the relevant variation in the first place. Others concern the problem of showing that the reproductive success of the individuals that we have reconstructed from the past really is lower than that of present animals, and if so, why. There is, fortunately, a short cut, a way round such practical obstacles.

Natural selection will, over the course of many years, have resulted in animals that appear to be 'designed' in certain ways – with eyes that give extremely good resolving power, for instance, or ways of catching their food that give a very high return of energy for the effort expended. Consequently, it may be possible to examine the animals to find out what they appear to be 'designed' to do and then use this to infer why it was that natural selection favoured them and eradicated their rivals.

To see what this might mean, we will look at a specific and very striking example: the echolocation system of insectivorous bats. These bats hunt small insects by sending out a brief pulse of sound and then listening to the echo that comes back reflected off the insect. The way that they do this is uncannily like that of the sonar or radar systems built by human engineers (Sales and Pye 1974; Simmons, Fenton and O'Farrell 1979).

The insects caught by the bats are so small that no echo would be reflected off them at all unless the wavelength of the sound used by the bats was roughly the same or shorter than the diameter of the insect. This means that the sounds have, of necessity, to be of a very high frequency – between 80 and 100 kHz to give a reasonable echo from an average-sized moth. Unfortunately, such ultra-high frequency sounds do not carry well over large distances because the sound energy in this range is rapidly absorbed by the atmosphere. The only way in which an echolocation system for detecting small objects could possibly work at anything over a few centimetres away from the sound source would be for that source to be extremely loud – loud enough to make up for the drop in intensity that inevitably occurs as the sound travels. The fact is that such bats as the little brown bat (*Myotis*) do use sounds of very high frequency (50 kHz and above) that are extremely loud (over 60 and sometimes as much as several thousand dynes per cm^2).

The bats listen to the faint returning echoes of these loud sounds in the split-second intervals of silence between their own vocalizations. Over a hundred times a second they may emit a pulse of sound and receive an echo back, decoding information about how far away the insect is, as well as what and where it is. If the system had been built by humans, we would say that it had been 'designed' to detect small moving objects. The use of very high frequency, very loud, rapidly repeated pulses of sound could hardly be bettered by a human engineer faced with the same task.

Now, this similarity between the ideal, man-made system and the actual behaviour of the bat can be used to give us yet another kind of evidence about adaptation. As with the other methods, we make a comparison, but here the comparison is made between a real animal and the most effective machine that human beings could design to do the job. The closer the match between a machine deliberately designed to perform a certain function and what an animal does, the more likely it becomes that we have identified the selection pressures which have been operating on the animal. In the case of the insectivorous bats, the close similarity between their echolocation system and the echo-detecting abilities of human sonar systems suggests that the bats' behaviour is adapted to catching insects on the wing and not, say, to scaring off predators or communicating with other bats. Death by starvation, through inability to catch food, would appear to have been the fate of the bats' now vanished rivals, not death through being eaten or death in any other guise. This conclusion about the adaptive significance of bat echolocation is reached without anybody ever having done a comparison of the reproductive success between a typical little brown bat and one of the same species that made an abnormally low frequency sound or one that was abnormally quiet. No-one has shown that the atypical bat would leave fewer offspring through not being able to catch so many insects. But having understood the 'design features' of the echolocation system makes this almost unnecessary. Catching prey is the only hypothesis about adaptation that adequately explains all the unusual features of the bats' behaviour. Another possible hypothesis, that the bats which make such sounds are safer from predation than bats which make other sounds, would not explain why the sounds that are made are at a frequency far above that which any possible predators can hear. Nor would it explain why the bats appear to adjust both the length and the timing of the sounds that they make so that the echoes of their own voices always come back in the intervals between the sounds they are making. Only the hypothesis that the sounds are an adaptation to the extraordinarily difficult task of catching small insects on the wing explains the complexity of bat calls. Only this hypothesis accounts for all such details as the duration of each sound, the way the sounds get shorter as the bat homes in on its prey, the frequency composition and the curious change in frequency (a drop of about an octave) during each pulse, and so on. The close similarity to a complex piece of human machinery allows us to reach this conclusion. The 'design features' of the machine and the parallels we see in the bat suggest that natural selection has discarded those alternative forms that did not live up to these criteria in the same way that an engineer might reject prototype machines that did not do what was required of them.

The value of this method of studying adaptation can be seen even more clearly if we now look at a negative example, that is one in which a hypothesis about adaptation has to be discarded in the face of 'design feature' evidence. It had been supposed for a number of years that one of the reasons why many fish travel in schools is that they gain a hydrodynamic advantage from doing so, that is they make use of the 'eddies' produced by other fish and so do not have to expend so much energy in swimming themselves. If fish schooling behaviour is 'designed' in this way, it should be possible to predict how the fish should position themselves to gain the greatest advantage from the swimming of their neighbours (Fig. 1). Partridge and Pitcher (1979) showed, however that for at least three species of schooling fish, the animals did not position themselves in this way. They may have been gaining some hydrodynamic benefit, but it was certainly not as much as they theoretically could have done by choosing different positions. The fish stayed too close to the fish in front to get the most benefit from the vortices they left behind (they should have stayed at least five tailbeats behind the pair in front) and not close enough to the fish on either side of them to get the most benefit from being able to push water against them (0.3–0.4 body lengths apart would have been the most effective).

These discrepancies in turn imply something about the selective pressures that have been operating on these fish. There might have been some selective advantage through energy saving during swimming, but it was almost certainly not as great as some other advantage – perhaps having a good view of approaching predators. Hydrodynamic advantages do not predict very accurately the structure of fish schools. Hence hydrodynamic advantages alone do not seem to have been the main selective force operating on the way fish position themselves with respect to one another.

The basic assumption behind this fourth method of studying adaptation is that by discovering what the 'design features' of animals are or are not, we have, by implication, specified the reason why selection picked out the animals that are now with us rather than their rivals that once existed but which, lacking this key design feature, failed in the struggle for existence. The rival alternatives are no longer with us in the flesh and we have to infer their existence from the intensity with which selection has apparently operated.

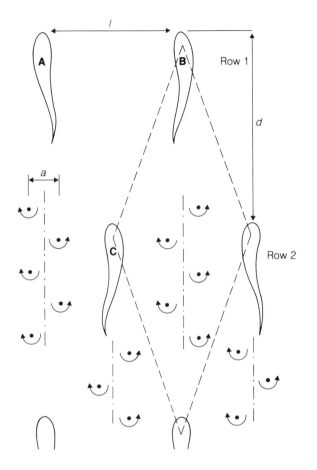

Fig. 1 Hypothetical fish school constructed on the assumption that fish position themselves so as to gain maximum energy saving from the swimming of other fish. Fish in the dotted diamond save energy in two ways: (1) Fish C, midway between A and B, receive induced veolocity from the spinning vortices shed by A and B. These vortices do not become stable immediately, so fish C should stay more than five tailbeats behind the pair in front (D). (2) Fish swimming close together can push off one another, thereby reducing energy expended. Maximum saving occurs at a separation of 0.3 body length (*l*). In three species of real fish – saithe (*Pollachius virens*), herring (*Clupea harengus*), and cod (*Gadus morhua*) – fish kept two to three times too far away from lateral neighbours as expected from the hypothetical school and swam too close to pairs of fish in front of them. (From Weihs 1975)

The more ruthless the selection has been, the more clear cut
will be the design features and the more animals that did not
possess them will have fallen by the wayside. Good design in
the few that remain can only have been achieved at the cost of
the many failures that were good at survival and reproduction,
but not quite good enough.

By finding out how good the design is in the animals that are
alive now, we get an idea of what selection has done in the
past. We infer the existence of the failures of the past from the
successes of the present. The ghosts of bats that could catch
insects but not particularly successfully are resurrected by our
understanding of where their shortcomings lay. And these
ghosts, these failures, reveal the adaptations of the animals we
now see around us. That is the idea behind using studies of
modern animals to draw conclusions about the long-term
actions of natural selection.

Conclusion

There are, then four well-worked out and tested methods for
studying adaptation. Two are what have been described as
'direct' because they involve trying to see death and des-
truction happening and to re-create the causes of past reproductive failure again in the present. The nature of the
adaptation is revealed when the rivals – real or artificial – are
actually seen to die or fail to reproduce. We know then the
reason why they died or failed to reproduce.

The other two, less-direct methods concentrate on the living
winners and try to infer from the characteristics that make
them successful what it was that gave their ancestors the edge
over long-dead rivals. One method compares species with each
other, searching for correlations with habitat and way of life.
The other compares animals with a hypothetical alternative –
a machine 'designed' to certain performance criteria. The
closer a real animal conforms to the same criteria, the more
confident we can be that natural selection, too, was favouring
animals that had these characteristics and not others.

Certainly 'adaptation' is a difficult concept which has been
much misunderstood. Certainly it is difficult to show precisely
how selection acted in the past. But the major pitfalls can be

avoided by remembering at all times that any adaptive hypothesis involves a comparison between two or more kinds of animals either real or hypothetical. This means that testing a hypothesis about adaptation must always be preceded by a specification of what were the alternatives available for natural selection to choose between. Having made it clear what the competitors were, we can then turn to the sorts of evidence we have discussed in this chapter to try and decide why some died and why some prospered. For it is the search for the causes of death and reproductive failure that is the essence of the study of adaptation. We can avoid confusion about it most easily by always remembering to see it in that unflattering and somewhat morbid light.

Further reading

Tinbergen's (1963) characteristically clear-sighted paper gives an excellent introduction to ethology in general and to the study of adaptation in particular. Gould and Lewontin (1979) criticize some of the ways in which adaptation has been assumed rather than demonstrated. Clutton-Brock and Harvey (1984) advocate the Comparative Method, Hoogland and Sherman (1976) show how advantages and disadvantages can be studied in the field, and Brown et al. (1982) give a neat example of an experimental approach to adaptation.

2
Optimality

Blue tit (*Parus caeruleus*). Photograph by Tony Allen.

Growing out of the idea of adaptation is the widespread belief that animals may not just be 'well designed' but that they may even be 'optimal'. Looking at the research papers published over the last ten years leaves the impression that almost everything that animals do has been described as 'optimal'. We find optimal foraging, optimal reproductive strategies, optimal times for threatening an opponent before retreating, and so on.

This emphasis on animals as 'optimizers' has led to an extraordinary degree of confusion about what 'optimal' really means and we cannot leave our study of adaptation until we have attempted to sort that out.

The two meanings of optimality

It is quite impossible to understand what 'optimal' means without realizing that it has not just one but two quite separate meanings. Firstly, there is 'optimal' meaning 'leaving the most viable offspring in a lifetime'. This can be called the long-term meaning and it is concerned with the reproductive success of an animal over its entire life compared to its rivals along the lines we were discussing in the last chapter.

Secondly, there is optimal in the short-term sense meaning that the animal appears to be optimizing some function in its day-to-day life, such as the amount of energy it is collecting in a certain amount of time. This is the sense in which the term is used in, say, 'optimal foraging'. The animal's ability to leave offspring may not be measured, but the animal is still described as 'optimal' if its behaviour conforms to some criterion, such as taking the most energy-efficient route to get to the most energy-rich food.

Of course, it is not strictly true to say that these two meanings are absolutely separate. Gathering food efficiently might well have a lot to do with leaving offspring – it could be the single most important factor contributing to it. But the connection is not straightforward. An animal that gathers food optimally might actually leave fewer offspring in its lifetime than an animal which gathers it less than optimally because it is so intent on feeding that it gets eaten by a predator. In other words, the long-term reproductive success kind of optimality and the short-term efficiency kind of optimality should be kept distinct. The thread that binds them, although it is undoubtedly there, may be a very tangled one indeed. Before we can unravel it, we have to look more closely at these two different senses in which 'optimal' can be used.

Long-term optimality

In the last chapter, we saw that our current picture of evolution is that, of the many mutants that have arisen in the past, the better ones prospered and it is their descendents that are with us today.

Now, for some people, this is equivalent to saying that the animals we see are the best of the ones that could ever exist, because given enough time, all possible mutations which could arise in a population will arise. Consequently, all possible mutations will have been given their chance and either failed in competition for some reason or demonstrated their success by leaving descendents which are still alive. The fact that a postulated mutant is not found in a population is, on this view, taken as evidence that it would not be successful because it has almost certainly arisen in the past, been given its chance, and failed.

This assumption is critically dependent on the further assumption that the environment has been stable for long enough to allow the slow processes of mutation and recombination to have come up with the best mutant the species is capable of producing. Hundreds and thousands of freaks and oddments, most of them less good at surviving and reproducing than normal members of the population, have to be born and try their luck. Only a few, a very few, will do better, and it may take a long time for these to appear.

There are, however, powerful reasons for suspecting that not many environments have been stable for long enough for that to have happened, long enough that is, for each possible mutation to have occurred at each site on the genome that could possibly affect the body of the animal. Certainly the broad general features of environments may well remain outwardly much the same for many generations. (Rain forests do not suddenly dry up.) But environments are not just physical things like temperature or rainfall. They also consist of other animals – competitors of the same species, predators and parasites out for a meal – each of them engaged in their own struggle for existence and each of them providing a constantly changing environment for the other animals in contact with them. Just as an animal has evolved an effective defence against a parasite, selection may then favour a new type of parasite that can get round the host's defences. Once this has happened, selection will then start favouring hosts that evolve new defences, and so on. Every time the optimum is approached, it slips away out of reach, with the parasite and its host engaged in what has been called a constant evolutionary dance, always on the move in evolutionary time. We might think, then, that we should define optimal in the long-term sense not as the best-possible mutant that could ever arise but more simply as the best of the mutants which have so far arisen in the present environment. This would leave open the question as to whether, with a bit more time or a change of environment, better animals than our present ones might arise. But this is not quite the same as the generally accepted meaning. More commonly, a statement that an animal is 'optimal' is taken to be a restatement of Darwin's theory of natural selection with a time element built into it. Optimality is a state which animals are said to approach when natural selection has been in operation for a certain length of time in the same environment. The possibility that parasites, predators, and rivals may see to it that the same environment rarely exists for very long is not taken as seriously detracting from the power of this formulation. It is assumed that the animals we now see are indeed optimal, subject to provisos such as these time-lag effects (Maynard Smith 1978; Dawkins 1982).

For the reasons we discussed in the last chapter, it is difficult, although not impossible, to study long-term optimality. Comparing the reproductive success of different kinds of animals may take years and generations, but at least in looking to the long term, we can be confident that we are dealing with the real stuff of evolution – with life and death, reproduction, and failure to reproduce. What, then, are we to make of optimality in the other, short-term sense? In experiments in which the reproductive success of an animal is not even measured, on what grounds can the animal be described as 'optimal'? The fact that large numbers of studies do conclude that animals are behaving optimally when those animals have not been allowed to breed at all must mean that 'optimal' is then being used in the somewhat different 'short-term' sense we have already mentioned.

Short-term optimality

To illustrate the meaning of short-term optimality, let us consider one particular example of how an animal has been thought to be optimal. Gaining food energy is vital for survival and one version of Optimal Foraging Theory (or OFT) suggests that animals should optimize the net amount of energy they obtain in a certain time. For example, a bird taking food from a small tree and gradually eating up all the food that is there would eventually come to a point when it would gain more energy if it left that tree and flew to another, where the food had not been depleted, even though the flight would itself use up valuable energy.

According to OFT, the bird should balance the energy gained in the new tree against the energy lost through flight. If this is correct and such a bird really is optimizing its net rate of energy gain, then it should be possible to predict precisely when the bird should leave one tree and fly to another, provided that the crucial variables – rate of gaining energy in the old tree, energy cost of flight between trees, and expected rate of gaining energy in the new tree – are known. It is like trying to work out what determines the route of a particular travelling salesman who is required to visit a number of different towns. If we have a hypothesis that in working out which order to visit them his only consideration is to cover the smallest number of miles, we can predict which route he should take. If he does take this route, even though it takes him much longer than if he chose a more circuitous but traffic-free road and means that he misses the best of the scenery, then we can have some confidence that we have discovered the true basis on which he makes his decisions.

Cowie (1977) used OFT to predict how long great tits (*Parus major*) should spend foraging in a given 'patch' (a sawdust-filled cup in which pieces of mealworm were hidden; Fig. 2) depending on the quality of the 'patch' (the number of mealworms in the cup) and the 'travel time' between patches. Knowing how much energy the birds used in moving between patches and the energy they gained through the food they were able to find, he could predict reasonably accurately when they should leave one patch and go on to another. This strongly suggests that the birds were optimizing net rate of energy intake.

Fig. 2 Great tit (*Parus major*) foraging in experimental 'patch'.
Photograph by Richard Cowie.

Interesting though such examples are, we must be careful not to
become sidetracked by the issue of whether the detailed predictions of OFT are borne out in practice (see Krebs and McCleery
1984 for a review). Our main purpose is to answer a different
question. *If* an animal optimizes energy intake in the short term,
does this tell us anything about its being optimal in the long-term
evolutionary sense? If it can be shown to forage optimally, in other
words, have we learnt anything about the way natural selection
acted on its ancestors in the past? And, conversely, if it does *not*
forage optimally, does this indicate that our ideas about the
adaptiveness of the behaviour were wrong after all?

The link between short-term and long-term optimality

In order to answer understand how short-term and long-term
optimality may be related, let us postulate a hypothetical

animal. This animal (a great tit, say) has been shown to be an optimal forager under a wide range of circumstances. It has been tested with a range of foods differing in energy content and in the time needed to search for and deal with them. It has been placed in conditions with long, energy-consuming journeys between one patch of food and the next and in others where the trip is short and easily accomplished. In all cases the bird behaves in a way which gives it the greatest amount of energy that it is theoretically possible for it to gain within the constraints of its own anatomy and physiology. Short of evolving a new more energy-conserving way of flying or a different kind of beak that enabled it to eat its food more quickly, there is no way for it to do any better. It is in this sense an optimal forager. What could we conclude from such a remarkable bird?

In the previous chapter, we encountered the idea that adaptation may be studied through the 'design features' of an animal's behaviour or morphology. If an animal closely resembles a human machine which has been 'designed' to certain criteria, this can be used to give an indication as to how natural selection has acted on the animal and what failures have been eliminated in the past. Inferences about adaptation are drawn even when no comparison is made between the reproductive success of different sorts of animal. The example we used was that of bat echolocation. Conclusions can be drawn about the adaptive significance of bat calls without any need to observe starvation in alternative sorts of bat. The 'design criteria' of the echolocation system were enough.

Could the same argument be used for optimal foraging? Could it be said that the optimally foraging animal is like a machine 'designed' to optimize energy intake? And if it is, could we not make inferences about what natural selection has done in exactly the same way as we did for the bats? It is certainly true that very few optimal foraging studies look at reproductive success (but see Drent and Daan 1980). In fact, no-one has shown that the lifetime production of offspring of an optimal forager is any greater than that of an animal which forages less than optimally. A comparable objection could be raised to the bats and yet we were still able to draw conclusions about their adaptiveness. The direct methods for studying adaptation – comparing reproductive success of alternatives – would be immensely difficult to apply in the case of foraging. But they may not even be necessary. If we have identified the 'design features' of the behaviour, perhaps this would be sufficient to tell us what it is adapted to.

It is very important to look at the logic of this argument because it is one which is assumed to be valid by a very large number of people. The supposition is that there is a direct connection between the short-term and the long-term meanings of optimality. In other words, if we can discover what an animal is optimizing in the short term, it is assumed we have a direct insight into why natural selection favoured its ancestors in the long term, and penalized the rivals of those ancestors.

As we saw in the first chapter, the aim of studying adaptation is to discover *why* some animals lived and others died and short-term optimality studies are thought to do just that. The unsuccessful animals might have failed because they could not get enough food or because they were eaten by predators or because they made poorly constructed nests that could not protect their eggs. There are many different possible reasons to account for the success of some and the failure of others. The goal of the study of adaptation is to find out what those reasons were – to elucidate the causes of death and reproductive failure.

The linking of short-term and long-term optimality assumes that those causes of failure long ago remain permanently impressed on the modern descendents of those who were not failures. Animals that were engaged in a struggle for existence which depended primarily, say, on getting enough food (in other words, their rivals were dying or failing to reproduce from starvation rather than anything else), would be favoured if, in their own lifetimes, they concentrated their efforts on getting as much food energy as they could. Those that were more efficient at gathering food than others would do better and the mutant that was best at food collecting would do best of all. This highly efficient food gatherer might not be very good at getting away from predators, but if predation were a minor risk compared to starvation, this would not matter. If starvation were the main arbiter between life and death, then the efficient food gatherer would flourish and leave many descendents. One that optimized energy intake would leave most of all. We now look at those descendents in our own time. We observe, let us imagine, that their entire behaviour appears to be 'designed' to maximize energy intake under all circumstances. Not having been present during the battle of their ancestors, we nevertheless infer that food was what the ancestral battle was about. The scars of that battle are still visible in their descendents.

That, at least in its simplest form, is the thread that binds the two sorts of optimality. The discovery of a 'design feature' or a function which is optimized in the short term is used as a way of recreating evolutionary events that happened a long time in the past.

The separation of short-term and long-term optimality

But what happens if the modern descendent does *not* forage optimally? Suppose, when we study its behaviour closely, we find that it is moderately good at collecting food but not really optimal. It wastes a lot of energy flying up into trees where there is no food. Can we conclude anything about its ancestors under these circumstances?

One obvious inference would be that getting enough food was not the only thing that made the difference between the success of its ancestors and the failure of its rivals. Perhaps going up into trees to look for predators also contributed, and the rivals that thought only of food (optimal foragers though they might have been) failed to leave as many offspring as the more wary animals.

There is a crucial point to be made here. The less-than-optimal forager that is aware of the approach of danger may, in the end, leave more offspring than the optimal forager whose eyes are never raised from its food. The optimal forager (optimality in the short-term sense) may therefore be less than optimal (long-term sense). This is not a contradiction. It forcibly illustrates the fact that short-term and long-term optimality are logically quite distinct from one another and should not be confused.

Because they are logically distinct, we should be careful not to put too much reliance on short-term optimality theories (such as optimal foraging theory) to tell us about long-term optimization. In particular, we should be wary of using measurements made in the artificial condition of the laboratory to draw conclusions about the adaptiveness of behaviour in the wild. A bird which is deprived of food before a test and then released in a laboratory environment in which there are no predators may be observed to forage optimally. But it does not follow that being better at gathering food was what gave its ancestors their main advantage over their rivals. It only suggests that when in danger of starvation and free from predation, they had an advantage through gathering food more efficiently.

If we did happen to find a real animal that took no notice of anything except acquiring energy under all circumstances, then the conclusion that feeding efficiency was the only selective force that had been at work might be quite reasonable. The chances are, of course, that such an animal is unlikely ever to be favoured by selection because in most environments, getting enough food is not the only criterion of success. Death comes in other forms besides starvation. Animals which are, on the average, best at evading all of these will become the progenitors of future generations. Their behaviour will be a compromise between getting enough to eat, avoiding being eaten themselves, mating, and so on.

So an animal may be 'optimal' in the long-term sense because selection will favour those that reach the 'best' compromise, but taken individually, none of its constituent behaviours may be optimal at all in the short term. Efficient food gathering may be compromised by the need to keep an eye out for predators. The best method for avoiding predators may be made impossible by the risks of dehydration or starvation they involve. The behaviour of the animal may, in fact, be an accurate reflection of the opposing selective forces that have acted on its ancestors in the past. But we should not be surprised if the picture we gain of what those selective forces were is not at all clear cut. It is only in the unlikely event of there being a single selective factor that caused death in some and favoured others that we would expect animals to optimize a single, simple function in the short term.

It is, therefore, a very difficult exercise, following the thread running from the behaviour of our present-day animals back into time to the selective forces which shaped their ancestors. As we have seen, it is not impossisble to follow it, tenuous and tangled though it often is. The way animals are 'designed' can be used to study adaptation but only if considerable care is exercised in interpreting the battles of the past from the survivors of the present.

The current use of the single word 'optimal' to cover both long-term and short-term processes is extremely confusing. It is essential to recognize the difference between short-term optimality and long-term optimality, for they are logically separate concepts. Only when the distinction between them is fully appreciated can the tangled thread be followed with any certainty.

Costs, benefits and economics

There are some more elegant and certainly more fashionable ways of expressing the ideas we have been discussing. Many biologists have recently started sounding a bit like economists. They use words like 'costs' and 'benefits' and even 'prices' and 'budget constraints'. Their papers have titles like 'The coal tit as a careful shopper' (Tullock 1971). They see an animal spending time on one activity or another much like a human consumer spending money on one commodity or another. Struck by the fact that economists assume that the organisms they are studying (in their case human consumers) are also optimizers, biologists have begun to phrase their ideas about evolution in economic terms. We will conclude this chapter with a brief look at some of them, but all we are doing is rephrasing and clarifying the ideas we have just discussed.

For economists, the function which is being optimized is known as 'utility', which is roughly the equivalent of 'satisfaction' or 'pleasure'. Human beings spend their limited incomes on the various options they can buy, and they are assumed to make choices (between buying cigarettes, say, or buying an exotic kind of food) in a way which gives them the greatest amount of this utility or satisfaction.

Biologists,   enthusiastically   embracing   the   economic

terminology, directly translate the idea of utility into what they see as its biological equivalent. And that is where the trouble starts, for, as we have seen, there are the two different ways that animals have been thought to optimize. Sometimes the biological equivalent of 'utility' seems to be taken as long-term reproductive success, with animals being said to optimize this and to pay 'cost' and 'benefits' in terms of loss and gain of viable children (e.g. Rubenstein 1980). At other times, it is taken as something more short term, like net energy intake per unit time. Here 'costs' means energy expenditure and benefit means energy intake. We have already seen that these different senses of optimizing have no simple relationship to one another. An energy 'cost' may well turn out to result in a 'cost' to offspring but not necessarily in any simple or one to one way. In other words, the animal which forages less efficiently may have fewer offspring, but, it may equally turn out to have more than the animal that forages more efficiently, through being warier about predators.

When applying the economic analogy, therefore, we must constantly keep in mind the distinctions of the previous section. McFarland and Houston (1981) have proposed some terminology to keep us on relatively firm ground. They suggest that we should take as the biological equivalent of utility (satisfaction) only the short-term optimality rules described by what they call the 'goal function'. 'Energy gained per unit time' would be an example of a goal function.

The longer-term reproductive success kind of optimizing, on the other hand, they suggest should be described as following a 'cost function'. If variation in the ability to get enough food is the only thing that contributes to the variance in reproductive success (the unlikely situation we imagined before), that would be equivalent to saying that unsuccessful animals died of not getting enough to eat and nothing else. The cost function would be an expression of this. A perfectly adapted animal in such circumstances would be one in which the goal function (short-term optimizing process) accurately reflected in the animal's day-to-day life this long-term need to get enough food; in other words, one in which the goal function, too, was geared exclusively to food finding. Cost function (long term) and goal function (short term) are thus assumed by McFarland and

Houston to be in close accord in animals in their natural environment, although separable in principle as we have already seen.

Many people are put off by the mathematics involved ('plant equations' make many of us think of lumbering vegetables). But the distinction made between different sorts of optimizing is exactly the one we have been groping for using the rather inelegant terms of 'optimizing in the long-term sense' or 'optimizing in the short-term sense'. The problem of the relationship between the two can be re-expressed as that between goal function and cost function.

We should not think, however, that just because we have found some mathematics to say what we wanted to say all along or that the ideas of optimality are also used by economists, that all the problems with it have disappeared. Optimality is still a difficult concept. Making the link between goal functions and cost functions is just as hazardous as trying to connect short-term with long-term optimality. Fashionable economic terminology should not make us lose sight of the genuine problems that are involved.

Further reading

Maynard Smith (1978) and Krebs and McCleery (1984) show what optimality studies try to achieve. Oster and Wilson (1978) apply optimality ideas to social insects, and McNeill Alexander (1982) to a broad range of structure and behaviour. Sibly (1983) shows how optimal group sizes in animals are inherently unstable.

3
Inclusive fitness

Hens and chicks (*Gallus gallus*). Photograph by Tony Allen.

Throughout the last two chapters, we have looked at natural selection in terms of the number of surviving offspring produced. We have stressed that what is important is an animal's lifetime production of offspring that themselves survive to reproductive age. This emphasis on the number of direct offspring as the sole measure of evolutionary success may have given rise to the feeling that our picture was incomplete. We have made no mention of relatives besides offspring such as siblings, nieces, or nephews. In view of importance attached to behaviour which benefits these other sorts of kin in much of the recent literature on behavioural ecology and animal behaviour, we surely should have done. We should, in other words, have taken into account other sorts of relatives, not just offspring, and included a measure of what Hamilton (1964) called 'inclusive fitness'.

Yes and no. Yes, aiding relatives other than offspring may be important. But no, this does not mean that counting only direct offspring is an invalid measure of evolutionary success. They amount to much the same thing.

This chapter will be about justifying this apparently paradoxical statement and showing that counting the number of offspring was in fact a central part of what Hamilton's much-cited paper was all about. Much cited it may have been, but, more to the point of this book, it has also been much misunderstood.

Aiding relatives and 'inclusive fitness'

Hamilton (1964) used the idea of 'inclusive fitness' as a way of calculating the conditions under which a gene might spread through a population, taking into account the effect that bearers of that gene might have on different sorts of relatives. In a moment we will see exactly how Hamilton defined 'inclusive fitness', which is what has been so greatly misunderstood. The misunderstandings that have arisen are not minor errors that can be quickly deleted from a few papers. They are (not to put too fine a point on it) major blunders. Grafen (1982) lists fourteen recent and commonly used textbooks on animal behaviour that either define inclusive fitness wrongly or fail to define it adequately. A quite extraordinary state of affairs! A central idea about animal behaviour misused and apparently misunderstood by almost everybody who discusses it. We are going to have to do quite a lot of unravelling. What follows is largely based on Grafen's 1982 and 1984 papers.

We start, where we should, with Hamilton (1964). Hamilton pointed out that there is nothing special, from a genetical point of view, about children. A child has a 50 per cent probability of sharing a given rare gene in common with its parent but then so it does with its siblings. More distant relatives also have a fair chance of sharing this gene, the chance being directly calculable from a knowledge of how closely related they are. This chance of sharing a rare gene is called the 'coefficient of relatedness'. Hamilton was not, of course, the first to realize this. But he was the first to provide a way of calculating whether a genetic tendency to help relatives would be successful enough to spread through a population. His purpose, then, was very similar to that of an 'ordinary' geneticist trying to calculate whether a gene for, say, white eyes in a fruitfly, would spread through a population of red-eyed flies. Geneticists talk about the fitness of a gene as its success relative to that of an alternative, in this case, the ratio of white-eyed genes to the genes giving rise to the normal wild-type red eyes.

Mutant genes for white eyes, however, have effects which extend no further than the body of the fly actually bearing them and which affect only that body's chance of leaving offspring. In the case of 'relative helping' genes, on the other hand, the effect is to help copies of themselves (and other genes) inside other bodies. These genes help the other bodies to have more offspring, which in turn will have a high chance of carrying the relative helping genes.

The calculation of whether such genes will spread is therefore quite complicated and has to take into account these 'other body' effects as well as the effects the genes have on the bodies that actually carry them. Hamilton coined the term 'inclusive fitness' to cover both these 'own body' and 'other body' effects.

As will become clearer as this chapter goes on, inclusive fitness is in one sense a measure of gene frequency (gene success relative to some standard) but it is expressed in terms of a property of individual animals. What is really going on Hamilton saw as selection at the gene level, but, if we like, we can pretend that selection is going on at the individual level by talking about the 'inclusive fitness' of individuals. If the pretence is not to lead us astray, the two must match. We should either be able to count gene frequencies or calculate inclusive fitness and they should give us the same answer. We carry on this pretence because it is so much easier for us to work with individual animals that actually walk around and do things than to think only in terms of shifting gene frequencies.

So, inclusive fitness is just ordinary, geneticists' fitness with room for effects on the reproductive success of relatives included but put in such a way that it can be applied to whole animals. Hamilton (1964) described it in this way:

> Inclusive fitness may be imagined as the personal fitness which an individual actually expresses in its production of adult offspring as it becomes after it has been first stripped and then augmented in a certain way. It is stripped of all components which can be considered as due to the individual's social environment, leaving the fitness which he would express if not exposed to any of the harms or benefits of that environment. This quantity is then augmented by certain fractions of the quantities of harms and benefits which the individual himself causes to the fitnesses of his neighbours. The fractions in question are simply the coefficients of relationship appropriate to the neighbours whom he affects: unity for clonal individuals, one-half for sibs, one quarter for half-sibs, one-eighth for cousins, . . . and finally zero for all neighbours whose relationship can be considered negligibly small.

The key to understanding what Hamilton meant, and why people have misunderstood him, lies in the curious business of 'stripping' and 'augmenting'. First, he says, inclusive fitness has to be 'stripped of all components which can be considered as due to the individual's social environment'. This means that we have somehow to work out how many children an animal would have had if he had no relatives around either to help him have children, or to hinder his rearing of children. Imagine, as an illustration, an animal that helps its brother to have children. This brother-helping animal sets up its territory near a brother. It rears some children of its own, but spends a lot of its time going next door to feed its nephews and nieces. This is time which it might have spent feeding its own children. It thus has fewer offspring of its own than it might have done, whereas the brother who is helped has more.

We have to try and work out the gains and losses resulting from this behaviour. According to Hamilton, we have to start by working out how many offspring an animal would have had by its own unaided efforts if it had been neither harmed nor helped by anyone else. We have to 'strip away' all the other offspring. We must ignore the ones that owe their existence to help from relatives and we must try to count the unborn ones that would have been alive but for hindrance from relatives.

The first part of this stripping is relatively straightforward. The offspring that were gained by the helped brother must be stripped away from him, because they would not be alive but for help from their uncle. We might even be able to get an estimate of how many of these extra offspring there were by comparing the reproductive success of nests with and without brothers that helped (cf. Brown et al. 1982).

The second part of the stripping operation is more difficult both conceptually and in practice. We want to know how many offspring the helper brother lost as the result of the effects from his social environment. This does not mean the number that he lost as the result of his own efforts at rearing nieces and nephews, but the number he lost *as the result of the actions of other animals.* For example, if his brother inveigled him next door and actually prevented him from having young of his own then this would be a 'harm from the social environment' and the offspring lost as a result of this should be stripped away. Stripping away unborn offspring means (through the double negative of removing something which has been taken away) effectively giving them back to the helper brother because we want to know how many offspring he would have had but for the interference from the helped brother.

The offspring that the helping brother lost as the result of his own actions in tending nieces and nephews are not part of this stripping operation because they are not casualties of the social environment. The helper brother was not harmed (he did not lose offspring) by the genotype of the brother to which he gave help, but he was harmed by his own genotype, which led him to give so much of his attention to his brother's children that he lost out on his own. We are stripping only those offspring that are affected – for better or worse – by variation in the genotypes of *other individuals* and trying to leave ourselves with an estimate of offspring that are affected only by variations in the genotype of the individual in question. We are concerned with offspring produced alone, unaided, and unhindered.

Secondly, we come to what Hamilton called 'augmenting'. To the offspring produced alone and unaided, we have to add relatives (devalued by the coefficient of relatedness) that are around because of the effects of relative-helping behaviour. Notice that this does not mean all relatives that exist. A solitary individual rearing offspring on its own without aiding its brother in any way is probably going to have nieces and nephews just because its brother (on the other side of the wood) is probably going to reproduce. Just having nieces and nephews is no real achievement. What matters, as far as the spread of brother-helping genes is concerned, are the *extra* ones an animal has as the result of helping. These are what could potentially give the evolutionary edge over the solitary kind of animal.

So we have to know how many extra offspring there are. In other words, we want the helped brother's 'stripped' offspring total first and then the augmenting effects of his helpful brother. Then we need the helper brother's stripped offspring total and have to 'augment' that, negatively, by the number of offspring he would have had if he had not been rearing nieces and nephews.

The purpose of the stripping and augmenting is to make absolutely certain that we have correctly tracked the changes in gene frequency. The big danger is to count the same individual twice and so to get an inflated idea of the frequency of the genes that it is carrying. For example, a single offspring of a brother which is helped could, unless we are very careful, be counted once as the son of his father and then again as the nephew of the uncle who reared him. It would look then as though there were two individuals where there is really only one and the result of 'brother-helping' would appear to have twice the genetic benefits that it really does. If a nephew owes his existence to the efforts of his uncle, then 'stripping' that nephew away from his real father makes quite certain that he does not get counted twice.

Unfortunately, there are many erroneous definitions of inclusive fitness that do inadvertently go in for double-counting and so give an inflated idea of the benefits of helping relatives. For example, Krebs and Davies (1981) say, in their definition of inclusive fitness, that it 'includes the animal plus 0.5 times its number of brothers and sisters plus 0.125 times its number of cousins and so on'. In other words, they are recommending the inclusion of *all* non-offspring relatives (suitably devalued), not just the extra ones that owe their existence to help from relatives as well as the inclusion of *all* offspring, not just the ones that the animal has by its own efforts. Grafen (1984) refers to this erroneous definition as the Simple Weighted Sum as it involves adding up all the relatives and devaluing them by their degree of relatedness.

If this definition is used to calculate the inclusive fitness of all animals that help relatives, then it will give the wrong answer. It will count many individuals twice or even many times over, as the offspring of the parent who had them and as kin of their relatives. This will give an exaggerated idea of the advantage of relative-helping, falsely inflating the sum of the relative-helping genes around.

If we keep our eye on the original, geneticists' idea of fitness as the number of copies of a gene in the population relative to the number of copies of its allele, we can see how very misleading this would be. We must not count a single copy of a gene twice just because it occurs in a single individual that could either be described as someone's son or as someone else's nephew. The son and the nephew are one and the same individual. He does not suddenly sprout more copies of his genes because he could also be described as the great-nephew of a third individual, the grandson of a fourth, and the cousin of a fifth and a sixth. In order to preserve the correspondence between gene frequency (gene-level way of calculating evolutionary success) and inclusive fitness (individual-level way of calculating the same thing), we must resist the temptation for all this multiple counting.

If an individual survives to adulthood, not all his relatives can take credit for him – he must not be included in all their inclusive fitnesses because when these are summed together to give the inclusive fitness of the relative-helping *genotype*, this will be wildly inflated. The only way to avoid this error is to ask firmly by whose efforts the individual managed to arrive at his adult state. If his parent, by his own effort, reared and cared for him, then he can be counted as part of his parent's inclusive fitness and should not be included in that of an uncle nesting many miles away who has had no effect on him. If he would have died but for the care of an uncle, however, he should be counted towards the uncle's total, not the parent's. Then we must not forget that the uncle, in order to rear this nephew, probably had to forgo at least some reproduction of his own and may even have suffered hindrance from his relatives. He may have gained some kin, but he has had to pay for them in the hardest currency of all – lost offspring of his own.

Offspring and inclusive fitness

Grafen (1982) documents a number of papers where inclusive fitness has been wrongly calculated. The commonest fault is to use a version of the simple weighted sum although, fortunately, the actual measurement of this in the field is so difficult that not too many man-hours have been lost through the pursuit of it. Only rarely is it known which animals are the cousins, nieces, nephews of which other ones.

But, if the simple weighted sum is difficult to measure, proper inclusive fitness in the sense that Hamilton defined it appears to be quite impossible. We seem to have to know not just what does happen but what would have happened under various unlikely circumstances such as if a socially living animal were suddenly stripped of all the harmful or beneficial effects of its social environment. We appear to have to measure not just how many offspring, nieces, nephews, and so on an animal has (difficult enough in itself) but actually how much it has contributed to the fact that they are alive and well and available to be counted.

Fortunately, there is an easy way out of this difficulty. Counting the number of offspring gives approximately the same answer. A bland statement to this effect appeared at the beginning of the chapter. It is now time to justify it and to see why something as down to earth and easy to measure as the number of children should give us the same answer as all those flights of fancy about stripping and augmenting. The link is as follows.

In a group of animals, all of whom carry the trait for relative-helping, there may be a number of offspring that owe their continuing survival wholly or partly to aid given by relatives other than their own parents. In an individual case, it is often very difficult, as we have seen, to know exactly how many offspring a parent would have had without help or hindrance from relatives, and how many children the relative would have had itself if it had not been helping or interfered with by the animals around it.

Now, 'inclusive fitness', like ordinary fitness, refers not to the reproductive output of a single individual, but to the success or otherwise of a whole genotype compared to some alternative genotype. An individual's success or failure, including the effect that he may have on relatives, is therefore important only in so far as it contributes to the average reproductive success of that genotype. And 'reproductive success' means adult offspring produced. By definition, the fitness of a genotype is increased relative to that of another if the number of copies of that genotype passing into the next generation is greater, and the parent–offspring line is the only way in which this passage can be made. Genotypes would go extinct if none of their representatives produced offspring. The same is true of the inclusive fitness of genotypes that help relatives other than offspring. The *average* inclusive fitness of all the relative-helpers is given simply by the number of adult offspring they produce compared with animals that do not aid relatives. The genotype as a whole does not aid non-offspring relatives as a substitute for rearing offspring. It gains its advantage by using this aid to produce more offspring. Some members of the genotype will have fewer offspring as a result of helping other individuals. But if the helped individuals have more than enough extra offspring in consequence to make up for this loss, the genotype will prosper. Its average fitness will be greater than the alternative if the average number of adult offspring produced by the genotype as a whole is higher than that produced by an alternative genotype in which no individual aids relatives other than its own offspring.

The number of adult offspring produced by a genotype will automatically reflect the extent to which relatives have helped as well as the extent to which their own reproduction has suffered. It will effectively have already been stripped and then augmented again by all the elusive things we find it so difficult to discover for a particular parent or a particular uncle. It will be the result of the net gain in adult offspring summed over all the individuals of the relative-helping type. As much help will have been received from relatives as will have been given to relatives, on the average for the genotype as a whole.

Consequently, the stripping we have to do for the average relative helper (calculating how many own offspring would exist in the absence of help or hindrance from other relatives) and the augmenting we have to do (adding in the harm or benefit done to relatives) exactly balance out when we do the sums for the whole genotype, leaving us with the number of adult offspring actually produced.

The importance of considering average inclusive fitness of the relative-helping genotype, rather than one individual's own inclusive fitness can be seen by considering an extreme case of relative-helping – the sterile castes of ants, bees, and wasps. A sterile social insect worker has no offspring – in this sense no fitness at all. But the average fitness of bearers of the conditional strategy gene that sometimes gives rise to queens and sometimes to sterile workers is very great, because many fertile offspring are produced as a result of this arrangement. We can legitimately ask how many fertile adult offspring are produced, on average, by bearers of this conditional trait and compare that with the average number produced by an alternative, such as bearers of a gene that always gives rise to fertile adults, never workers.

Sometimes more fertile children can be produced by having some sterile children who are helpers than by making all children into fertile ones. The actual number produced is despite loss of reproduction by the sterile workers and because of aid given by them. We do not have to go through the stripping and augmenting calculations because the net result of all the attendant costs and benefits will be reflected in reproductive output. If we count the number of fertile offspring of the two types (those with and without sterile offspring), this will tell us whether the average inclusive fitness of the one is greater or smaller than that of the other.

It should perhaps be pointed out that although comparing the number of offspring in this way does give a valid indication of the comparative evolutionary success of two genotypes, the average number of offspring does not exactly equal the average inclusive fitness to be found in Hamilton's equations. The reasons for this are technical and need not concern us here (Grafen 1984), and they do not invalidate the idea of comparing offspring numbers rather than inclusive fitnesses. There may also sometimes be problems with counting offspring if it is not possible to identify which of them carry which genes (Grafen 1982). It is then impossible to work out whether a given genotype is at an advantage or not simply by counting offspring. Provided it is possible to identify genotypes, however, this problem does not arise. The number of offspring is what really matters.

All this is, of course, most extraordinarily convenient. It is far easier, in practice, to count numbers of offspring than it is to count all the relatives that are demanded by the simple weighted sum. Yet a comparison of offspring numbers is a much better measure of evolutionary success as well as being closer to Hamilton's original idea 'of inclusive fitness'.

It might be thought, though, that what we have in fact done is to show that inclusive fitness is not important after all. If we are back at measuring the number of offspring, what is the point of it? Why should we be concerned with the effects of non-offspring relatives when offspring relatives seem to be the only ones that count? This is not what we should conclude as we can see from looking closely at what Hamilton did show.

Hamilton demonstrated how a trait could spread through its effects on the reproductive success of animals other than direct offspring. He emphasized that this was not a new sort of selection process but one which operated in essentially the same way as ordinary 'parental care' selection. In both cases the 'costs' and 'benefits' are paid in the same currency – loss and gain of adult offspring. Even with parental care, there is cost (to future offspring) of nurturing present ones.

But genotypes that do not produce adult offspring cannot be successful. Hamilton showed that one of the ways that genotypes can increase their production of adult offspring is to have some members of that genotype diverting their efforts from reproducing themselves to helping the reproduction of other members of the same genotype. Wanting to express this idea not in terms of genotypes and gene frequencies but in terms that could be applied to individual animals, he devised 'inclusive fitness'.

Whether a genotype can increase its production of adult offspring in this way will depend on various contingencies. If one brother has a very good territory and is good at reproducing, and a second is underweight and has a poor territory, more adult offspring may be reared if the second weaker brother spends his time chasing predators away from nieces and nephews than if he attempts to rear a brood of his own. The genes for helping brothers under such circumstances will spread. The fact that the criterion for whether or not this is the best strategy is how many adult offspring are produced emphasizes the importance of looking at inclusive fitness. Raising offspring to reproductive adulthood is what is favoured by natural selection. Aiding relatives is one way in which this can be achieved.

The concept of inclusive fitness is quite fundamental to our understanding of whether a given trait, particularly a behavioural one, will be successful in evolutionary terms. Helping a relative other than a child can be adaptive in some circumstances. But it is not without cost and Hamilton's formulation clearly lays out the conditions under which we may expect it to evolve. For theoretical purposes, his equations with their stripping and augmenting requirements are sometimes necessary (Grafen 1982; Maynard Smith 1982a) and helpful. But they also show that, when it comes to studying real animals, approximately the same result can be achieved by comparing numbers of adult offspring. This, fortunately, is relatively easy to measure in the field. It is also, no more and no less, what evolution is all about.

Further reading

It is best not to read too much about inclusive fitness as much of what has been written is misleading. The papers by Hamilton (1964) and Grafen (1982,1984), read carefully, will set you on the right lines.

4
Genes and behaviour

Maras (*Dolichotis patagonum*). Photograph by Andrew Taber.

A consistent theme of this book so far has been that many of the confusions surrounding the evolution of behaviour can be avoided by constantly reminding ourselves that natural selection is about competition between two genetically different alternatives. Whenever we are confronted with a problem in adaptation, we have to ask what those alternatives were and why one did better in competition than the other. Unfortunately, our troubles do not end there because those questions, innocuous though they may sound, have themselves aroused intense opposition. They have given rise to accusations of 'genetic determinism' (Gould 1978; Rose 1978). The idea that behaviour has evolved and that there might be genes for behaviour in the way that there are genes for other characteristics is what seems to be objected to. Postulating 'genes for' behaviour is taken as synonymous with the view that genes control all we do.

It is impossible to dismiss this opposition by the simple, if correct, reply that this is not at all what is implied by the theory of natural selection. The criticisms have gone on for so long and have such a hold over many peoples' imaginations that a more explicit rebuttal is called for. In this chapter, we will try to tackle that task. But to explain all that has to be explained, we have to unravel a number of other problems first.

Before being able to see why postulating genes for behaviour does not in fact imply that genes control or dominate that behaviour in any sinister way, we have to look a bit more closely at what we mean by 'a gene for behaviour'. Then, having cleared up this definitional problem, we have to look at the evidence for the existence of such genes to see whether it is true that they possess the sinister, dominating properties that have been attributed to them. Then, equipped with a definition and some evidence, we will finally be in a position to separate clearly the role of genes in natural selection and their un-founded reputation as dominators or dictators of what we do.

'Genes for' behaviour

Adaptation, we saw, implied that natural selection has favoured one type of animal over another of a different colour, shape, or behaviour. If we wish to study adaptation, therefore, we have to search for these different types of animals and see why one did better than the other. Very often the original types are not available for us to study for the simple reason that natural selection has already occurred and eliminated all but a few. Studying adaptation often means adopting less-direct approaches as we saw in Chapter 1, but all approaches involve making some kind of comparison. We compare the evolutionary success of one kind of animal with that of another.

Comparison dogs all studies of adaptation. The word implies comparison – of something, with something. Natural selection cannot operate unless animals show differences between which selection can occur. And we can go further: natural selection cannot work unless those differences are inherited, unless, that is, the traits which gave their parents the edge over their rivals are handed on to their offspring.

Natural selection, then, implies not just variation but genetic variation of some sort, however slight. This is not to say that this genetic variation exists now. Natural selection may have eliminated the variation that once existed. But we assume that some genetic variation must have existed in the past. There must once have existed black-headed gulls that did not remove eggshells and lost their chicks to predators as a consequence, even if such gulls are not to be found nowadays. We assume that these gulls must have been genetically different from the more successful eggshell-removing types. Otherwise we would be wrong to call eggshell removal an adaptation or to imply that natural selection had been at work.

So, a belief that behaviour has evolved in this way does indeed commit us to an assumption about genetic variation. We assume that there either are or there have been in the past individuals that differed from one another genetically with respect to behaviour. It is in this sense, and in this sense only, that we use the phrase 'gene for' behaviour. Just as 'adaptation' is correctly rendered as 'adaptive difference' (between two types), so 'gene for' can be replaced by 'genetic difference' (also between two types).

All that is implied by using a phrase like 'gene for helping a relative' is that there is some genetic difference between animals that do help relatives and those that do not, a slightly greater tendency to regurgitate food, for instance, when arriving at the nest. It does not mean that there is a single gene which, by itself, controls all the animals' interactions with relatives. It simply means that a small genetic difference could be responsible for making an animal into a good regurgitator or a bad regurgitator, perhaps through affecting the level of hormone secreted by a particular gland.

This is not, incidentally, in a new meaning of a 'gene for' something. It is precisely what 'ordinary' geneticists, studying eye colour in fruitflies, height in pea plants or inherited blood diseases, also mean by it. The difference between a person with normal red blood cells and one with the disease called sickle-cell anaemia has been traced to a difference in a single amino acid in the haemoglobin chains of the red blood cells. So the 'gene for' this serious and often fatal disease could be said to be the gene that alters a single polypeptide chain in the haemoglobin molecules. This gene for sickle-cell haemoglobin is not responsible for the production of the entire red blood cell or even the whole haemoglobin molecule (the normal and the affected person will have most of their genes for this in common) but it is responsible for the *difference* between normal and sickle-cell individuals.

'Genes for' white eyes in fruitflies or wrinkled seeds in pea plants refer to similar sorts of difference. In neither case is a single gene seen as giving rise to a whole eye or a whole seed – many genes are involved in both cases. But one or a few genes can be responsible for a difference between different types and in that restricted sense, can be regarded as a gene or genes 'for' those characters.

This implies that there exist, somewhere in the DNA molecules, changes that bring about alterations in the structure of the animal bearing them: not, when put in this way, such an outrageous claim. We know that genetic variation exists and has been documented for a large number of morphological characters. We are quite used to it for blood groups, coat colour, milk yield, and a host of other things. Is there also evidence for genetic variation in behaviour?

The genetics of behaviour

There is, in fact, very good evidence now of this, of a sort very similar to that previously obtained for morphological characters. This is hardly surprising. Behaviour results from the interaction of sense organs, nervous system, muscles, and other parts of the animal's body. Variation in any of these might be expected to affect behaviour as well. The cell membrane of the tiny single-celled *Paramecium*, for instance, is affected by the presence of mutant genes. Animals with abnormal cell membranes in turn have strange behaviour. For example, the 'Pawn' mutant is unable to go backwards because its cell membrane does not allow calcium ions through in the normal way. The electrical spike produced by positively charged ions rushing into the cell is unable to occur as it would in a normal animal. This means that the cilia on the outside of the membrane cannot change their direction of beating and the behavioural disorder of never being able to go backwards results (Kung et al. 1975).

In male crickets (*Teleogryllus*) which attract females by singing (produced by scraping one wing rapidly over another), an alteration in the pattern of nerve impulses going to the wing muscles completely changes the nature of the song. Too many or too few nerve impulses or even just the right number of nerve impulses at the wrong times makes the wings produce a song that female crickets simply do not find so attractive (Bentley and Hoy 1972).

Genes in the male are responsible for affecting the number and patterning of nerve impulses that go to the wings. Crossing two species that have different songs results in males that have songs intermediate between the two. The pattern of nerve impulses to the wings as they sing is also intermediate. So, genes affect nerve cells which affect wing muscles which affect the song produced which in turn affect the attractiveness of the male to females. The route from genes to behaviour is long but it is traceable, at least in this case.

In other species of cricket, the amount of singing a male does is also affected by his genes. In a species called *Gryllus*, Cade (1981) showed that there were two sorts of male. Caller males sit and sing for females. Satellite males sing much less but intercept the females on their way to the singers. Caller males have the benefit that the females know where they are but are much more likely to be parasitized because their parasites, like the females, use the song to locate them. Satellite males with the advantage of being less parasitized coexist side by side with singers, the balance of advantages and disadvantages between them being apparently equal.

Breeding experiments show that the difference between callers and satellites appears to be genetic. Breeding from caller males gives rise to sons that sing a great deal, whereas breeding from satellite males gives sons which tend to be silent for most of each night.

In many of the genetic studies of behaviour, however, the situation is not as simple as appears from these examples. There is often an additional complication in that the environment can be shown to contribute substantially to the observed variation between individuals. In other words, how the animals were treated in early life or the environment they find themselves in as adults affects their behaviour just as much as what genetic strain they belong to.

For example, some strains of mice are much more aggressive than others. Some mice will hardly ever chase or fight each other, whereas others will be much more ready to do so. Much of this variation is genetic. Even if mice from an unaggressive strain are raised by mothers from an aggressive strain, they still tend to fight like their real, genetic parents do (Southwick 1968). At the same time, the experience of the mice when they are young also matters. Even within one strain, there can be more or less aggressive individuals depending on what they have experienced during their own lifetimes (Lagerspetz 1969).

It is important to realize, however, that the existence of environmental effects does not invalidate the idea that some of the differences between individuals may still be genetic. The fact that many genes may be involved and that the environment may play a role too, simply means that we have to employ rather more sophisticated ways of studying the genetic effects (Ehrman and Parsons 1976).

The reason for this is that natural selection does not demand that *all* variation should be genetic. What is needed for it to operate is that at least some part of the variation between individuals should be genetic. The more of the variation that is genetic, the faster it will operate, but even a small genetic difference between individuals will in the long run be enough. The course of evolution can be altered as long as successful individuals can pass on at least some of their characteristics of success.

Many behavioural characters have now been shown to exhibit this genetic variation. Even something as complex as 'recognizing relatives' may have a genetic base. Sweat bees (*Lasioglossum*) live in burrows in the soil with the entrance to the burrow patrolled by 'guard bees', which allow only certain bees in to the nest. Whether or not the guard bees allow another bee in is, quite remarkably, directly correlated to the closeness of their relationship with it. If a full sister tries to get in, she will almost certainly be allowed to, but aunts, nieces, and first cousins are likely to be prevented in direct proportion to their degree of relatedness. Even sisters that the guard bee has never met are allowed in according to their genetic closeness with the guard bee. Half sisters with a different father are less acceptable than full sisters with the same father. These in turn are less acceptable than full sisters of parents that are themselves related and which therefore have an exceptionally close genetic affinity (Greenberg 1979).

It seems likely in this case that genetic differences between individuals result in their having slightly different odours – the more closely related the bees are, the more similar they will smell. The guard bees perform their relative-excluding feat by turning away bees that have the wrong odour. The recognition of relatives is accomplished by a subtle mixture of genetic variation (in odour production) and learning (of familiar odours).

From these and other examples, we can see that it is quite plausible to postulate genetic variation in behaviour for a number of different traits, including complex ones such as interactions with relatives. There clearly are genetic differences between animals in the way they behave and in this sense, there are many 'genes for' behaviour.

Genes, behaviour, and genetic determinism

Does the existence of genetic differences in (genes for) behaviour imply that those differences are in some way fixed and impossible to get rid of? In many people's minds, this is what the very existence of genetic variation implies. Genetic traits are seen as ones that no-one can do anything about. To concede that genes affect behaviour becomes equivalent to saying that genes determine behaviour. They become the controllers, dictating what animals do at every stage of their lives. The fallaciousness of this view can be seen by considering two examples.

There is a serious single-gene inherited disease called

phenylketonuria. Children inheriting a copy of this gene from both of their parents are severely mentally retarded and usually die before reaching their teens. The gene works by preventing the formation of an enzyme, which, in normal people, converts one amino acid, phenylalanine, into another (tyrosine).

Phenylanaline accumulates in the brains of children with phenylketonuria and causes the tragic effects on their mental development. The trait is due to a single recessive gene and we even know the biochemical effects that it has on the body. It is, however, quite possible, to mitigate its effects by means of relatively simple changes in the environment. If the disease is detected early enough in life, affected children can be fed on a diet containing no phenylalanine. The result is that phenylalanine does not accumulate in the brain and the children are normal or nearly so (Hsia 1970). A genetic difference between normal children and children with phenylketonuria can thus be largely eliminated by manipulation of their environment. The children still have the genes for phenylketonuria, of course, but the effects of the genes are overcome.

A second example that illustrates much the same point concerns the ability of mice to run through a maze to find food. The particular 'maze' concerned consisted of a series of ladders, tightropes, and so on, which mice of different genetic strains had to learn to negotiate in order to get food. Mice of some strains were much quicker at learning this task than mice from other strains, but the differences appeared to be largely environmental (Henderson 1970).

This initial conclusion was arrived at by comparing mice all reared in the same standard laboratory mouse cages. When the experiment was repeated, with the mice reared in 'enriched' environments (cages with toys and things for them to climb on), the learning ability of all the mice improved. An enriched environment in early life seemed to make them all better able to cope with the maze. But what was particularly interesting was that the genetic differences between mice from different strains became more apparent too. When the mice were reared in enriched environments, mice from one strain showed themselves to be very much quicker at learning than mice from the other strain and this could be shown to be a genetic difference. This difference had not been apparent when mice of both strains had been reared in ordinary environments. Neither strain was particularly good at learning. It needed the enriched early environments to sort them out.

These two examples serve to emphasize that genetic differences may show themselves under some circumstances but not in others. Genetic differences between animals may not always be apparent. They may need a particular environment to show themselves at all. They are certainly not fixed and impossible to eradicate. Once again, this is not a peculiarity of the genetics of behaviour. It is equally true of 'ordinary' genetics. Suppose we had a strain of plants, all genetically much the same, but we put them into different sorts of soil – some with rich fertilizer, some with poor stony sand. Almost certainly the plants, even though they were genetically identical, would grow to different heights depending on the nourishment they received. Most of the variation in height would be environmentally produced. Then suppose we took two genetically different strains of the plant and grew both of them in the same soil, keeping the environment as similar as possible. Now, any variation in plant height would be largely genetic – and the environmental contribution rather small.

Clearly, the extent to which differences in the heights of plants can be described as genetic will depend on exactly how the experiment is done. The question 'are differences in plant height genetic?' has no single answer. Genetic differences may be very obvious if two strains are grown in the same soil, but quite overwhelmed if some plants are given an especially favourable environment to grow in. A plant from a 'genetically short' strain grown with extra fertilizer might even grow taller than one from a 'genetically tall' strain grown in poor soil.

All these examples show that there is nothing immutable or inevitable about genetic differences. Genetic differences may be radically altered, mitigated, or even reversed by altering the environment. Genes contribute to the observed differences between individuals, in their behaviour as in other things, but their contribution is not sacrosanct. It can be lessened or enhanced in just the same way as the contribution from the environment can.

That should be a relief to anyone who has misunderstood this fundamental point. It is possible for us to believe in adaptation with its implication of underlying genetic variability without having to become genetic determinists. We do not have to believe in the dominion or dictatorship of the genes nor to see natural selection making animals (ourselves included) into puppets, manipulated by their genetic masters within.

Genetic variability is indeed a logical necessity for the theory of natural selection, and there is overwhelming evidence for its existence even though the details may not be known as fully in a given case as we might like. But genetic variability is nothing sinister, even though it has often been construed in that way. Such constructions are based, as we have seen, on a misunderstanding of what the science of genetics is about and what a geneticist means when he talks about a 'gene for' something.

This is not the whole story, however. The arguments about the relative importance of genes and environment in behaviour cannot be so lightly dismissed. It would take more than a discussion of simple genetics to resolve finally what has been called the 'nature–nuture' argument. People have argued for many years about whether any behaviour can be described as 'innate' or 'instinctive'. This whole issue is so deeply rooted in the history of ethology that in order to understand what is involved and how it relates to the present discussion of 'genes for' behaviour, we must temporarily go back in time. The next chapter will involve a bit of history.

Further reading

Ehrman and Parsons (1976) and Partridge (1983) give good coverage of behaviour genetics. Dawkins (1982, ch. 2) describes the distinction between the role of genes in natural selection and genetic determinism. Anyone who believes that the arguments about genetic determinism have been exaggerated should read Rose, Kamin and Lewontin (1984).

5
Innate Behaviour

Reed warbler (*Acrocephalus scirpaceus*) feeding cuckoo (*Cuculus canorus*). Photograph by John Horsfall.

Of all the controversies that have arisen in biology over recent years, few have generated so much misunderstanding as that of whether any behaviour can or should be described as 'innate'. This is very largely because the word 'innate' itself has been interpreted in a number of different ways and each of its different meanings has aroused controversies of its own. In this chapter, we will try to make some sense of the confusion by looking at each of the meanings in turn and trying to see the connection, where there is one, between them.

Innate as 'genetic'

Firstly, there is innate meaning the same as 'genetic'. The dictionary definition of innate is 'inborn' so it is intuitively reasonable to interpret this as meaning 'in the genes'. The last chapter showed us, however, how very careful we have to be with the relationship between genes and behaviour. To talk about 'genes for' behaviour implies that there are genetic differences between one variety of animal and another. It does not imply that the genes are the only cause of the behaviour or that the genetic differences are fixed and immutable.

If we substitute 'innate differences' for 'genetic differences', then the same arguments apply. It is quite possible for differences between individuals to be genetic or innate in origin, as the examples in the previous chapter showed. But it is a fallacy to think that those differences cannot be changed by the environment. The differences may lie 'in the genes' but they are then no more fixed than if they were 'in the environment'.

The controversies over this meaning of innate are thus, by definition, the same as those for 'genetic' which we discussed in the last chapter. Many of these, as we saw, arise from a misunderstanding of what geneticists mean when they talk about a 'gene for' a character. They are referring to a difference between animals that arises from differences in their genotypes. Clearly, this only makes sense if they are talking about more than one individual. There cannot be a difference between two animals if there is only one animal. Genetic differences or genes for behaviour are thus properties of populations or groups and cannot be applied to single animals.

Because 'innate' is often applied to the development of behaviour in one single animal, it must then be being used in a different way altogether.

Innate as 'the opposite of learnt'

The argument about the 'innateness' of development of behaviour in the individual has been going strong now for nearly fifty years. It centres on whether it is possible to say that any behaviour is truly 'innate'. Konrad Lorenz (1932, 1937) was firmly of the opinion that it was. He saw an animal's behaviour as divided up into 'instinctive behaviour patterns', which he believed to be based on inherited nervous pathways, much like reflexes only more complicated. These instinctive behaviour patterns (*Instinkthandlungen*) were 'innate' or 'inborn'. An animal did not have to learn how to do them either by imitation of another animal or by trial and error on its own. It certainly did not have any kind of 'insight' into what it was doing.

In Lorenz's view, there was a clear-cut distinction to be made between innate and learned behaviour, which, in practice, it was possible to make by rearing a young animal in isolation (so that it could not copy another member of its species) and denying it any opportunity for learning on its own. A stickleback reared in isolation and performing the zig-zag courtship dance just like any other male stickleback the first time it saw a female or even a model of a female, would be an example of 'innate' behaviour – the genes telling the animal what to do.

Even at this stage, the seeds of confusion had been planted. Lorenz equated 'innate' with 'inherited', which is not very surprising, but he also equated 'all effects of the environment' with 'learning', which is not everyone's idea of what learning means. The flight of fruitflies is affected by the temperature they experience during the early stages of development because this affects whether the balancing organs develop into wings (Strickberger 1968). This is an effect of the environment, but it is stretching the definition beyond recognition to call this 'learning'.

Nevertheless, Lorenz persisted in this view. He repeatedly asserted that there are only two types of factor that affect the development of behaviour – genetic ones and learning. By devising protocols that excluded the effects of learning (imitation, conditioning, etc.), he believed that he had demonstrated the importance of genetic factors. If the behaviour was not learnt, in other words, it must be innate. Exclude the one and only the other is left.

Lorenz's views of innate behaviour grew out of his many years of watching animals grow up and perform complicated and appropriate behaviour with no opportunity to learn what to do. They were published in English in 1950 and were criticized mercilessly from both sides of the Atlantic, but particularly from America.

How could Lorenz be so naive, wrote the critics (e.g. Lehrman 1953) to believe that behaviour can be neatly packaged into 'innate' and 'learned' components quite separate from one another? What about other environmental effects besides 'learning' (such as the temperature one we just mentioned)? What about subtle learning effects an experimenter might not be aware of? And, above all, how could anything be entirely innate, entirely determined by genes? Every animal needs an environment to develop in. At the very least it needs food, oxygen, and a certain range of temperatures. If these are absent, it cannot develop and there is no behaviour. The idea of any behaviour being 'innate' – entirely within the genome and unaffected by any environmental factors – was dismissed as quite meaningless.

There is an important sense in which these criticisms are valid. Whereas it does make sense to say that differences *between* individuals are entirely genetic, it does not make sense to say that the development of behaviour *within one* individual is entirely genetic. Genes need environments in order to build bodies which can then behave. All behaviour is therefore due to both genetic and environmental factors. That is a truism. It is hard to believe that the arguments had no more substance than a denial of this obvious, self-evident fact. In fact they had, and the controversies flared largely because of misunderstandings, on both sides, as to what 'innate' actually means.

Accepting the obvious fact that all behaviour is due to both environmental and genetic factors, there are two possible ways of still keeping the idea that some behaviour is 'innate'. There is some behaviour that can be called 'developmentally fixed' – all animals of a species do it in almost any environment they have been brought up in. Short of boiling the animal or putting it in an environment in which it simply cannot survive, we find that almost no change in the environment has any effect. The animal grows up and carries out the behaviour.

A good example of 'developmentally fixed behaviour' is shown by the *Teleogryllus* crickets we came across in the last chapter. The males sing in a way that is characteristic of their species and different from any other species. Even massive environmental changes – rearing crickets in isolation, subjecting them to the songs of other species, and so on – have no effect. The male cricket persists in singing his own species' song. Of course, in the future it may be discovered that there is some environment that does alter the song. Perhaps eating a particular sort of yellow fruit will have the effect of making the male sing a different sort of song. But that has so far not been discovered. The singing appears to be very 'developmentally fixed' and certainly not learnt from other crickets (Bentley and Hoy 1972).

We could, on this definition, refer to the singing as 'innate', but it would be necessary to point out that this did not mean that the environment was not important. Of course, developing male crickets have to eat and breathe and have hundreds of other interactions with their environments. But it looks as though, within the range of viable environments for that species, precisely which of those environments the crickets happen to grow up in has no effect on the singing.

In this respect, the behaviour is different from, say, human speech, which is highly dependent on the precise acoustic environment a person grows up in. So, although it might perhaps be useful to put a little flag 'innate' on the cricket singing to indicate its developmental fixity, we might also have to be ready with a lot of disclaimers for people who thought we were denying any role for the environment at all.

There is, however, another possible meaning for 'innate', which is rather more subtle and particularly important for any understanding of the controversies surrounding the word because it is the one that Lorenz himself put forward in a revised (or clarified) form of his ideas in 1965. In the light of the many criticisms of his earlier papers, Lorenz reiterated what he saw to be the central point about the development of species-typical behaviour: that it is adaptive, that the animal in some sense 'knows' what to do without having to learn. The male stickleback 'knows' that rival males are red underneath; in Lorenz's view the important thing is to understand how it comes by this information. It might learn it or it may carry it in its genes. Lorenz saw no other way in which the animal could come by the information. The fact that the animal needs to have many interactions with its environment (food, oxygen, etc.) in order to develop the behaviour Lorenz saw as quite irrelevant because no amount of food or oxygen could give it the *information* that rival male sticklebacks are red underneath.

To decide whether aggression in male sticklebacks should be called innate or not, Lorenz therefore thought it was quite unnecessary to investigate the possible role of all environmental factors that could be thought of. It was necessary only to discover how the information 'rival males are red' enters the stickleback.

Rearing the fish in isolation from other members of its species (making sure that it cannot see its own reflection in the sides of a glass tank) and seeing how it performs the first time it ever sees another male stickleback would go a long way to settling this question. Attention is thus fixed on the role of particular environmental factors (in this case, exposure to fish of the same species) rather than on the whole range of environmental factors of the other, 'developmentally fixed' method. The only thing that matters here is whether the development of the behaviour is affected by alterations in the one crucial factor that could give information about what rivals look like. The cricket singing would also be described as 'innate' in this sense, but on the basis of the one finding that males 'know' what to sing without having to learn through hearing other crickets.

This view of Lorenz's has immense appeal. He wanted to emphasize the fact that so much of what animals do is adaptive the first time they encounter a particular situation. The world of animals is not a gigantic Skinner Box in which they gradually learn, by trial and error, what to do and what not to do. They come into the world 'equipped by nature' to behave in ways that are likely to help them survive and reproduce. And this, Lorenz maintains, demands an explanation.

There is an intuitive sense in which 'innate behaviour' is a special category. If an animal 'knows' what its rival, mate, or food looks like without previous experience, it is showing a different sort of behaviour from having to learn by being rewarded or punished. Or is it? Are the two sorts really so different? Even 'learning' is not writing on a blank slate. Animals are much better at learning some things than others. Chaffinches have to learn their song from other chaffinches in order to produce the full species-typical sound. But they will only learn certain sorts of sound – chaffinch song or elements of it (Thorpe 1961). A chaffinch cannot be taught to sound like a canary even though it may have to learn to sound like a chaffinch.

Such 'constraints on learning' chip away at the distinction we might intuitively want to make between 'innate' and learned behaviour. There are also examples of 'innate' behaviour being modified. Laughing gull chicks have an 'innate' tendency to peck at their parent's bill in order to make them regurgitate food. They do not need to learn. They emerge from the egg with 'information' about what a parent gull looks like even though the information is rather crude. They will beg particularly vigorously to long thin red knitting needles because these have many of the characteristics of the parent's bill (Hailman 1967). The initial innate response, however, is subsequently modified by experience. The chick learns, as it grows, the finer aspects of what its parent looks like and becomes much more discriminating in what it shows the begging response to (Hailman 1967). Older chicks will beg only from real gulls or models that look very like the parent laughing gull.

The distinction between 'innate' and 'learned' is further eroded by the song sparrow. Song sparrows reared in isolation develop perfectly normal song-sparrow song. But song sparrows deafened so that they cannot hear themselves singing, develop only a very rudimentary type of song. The birds need to hear themselves singing (Mulligan 1966). They practice until the song sounds right, 'right' meaning that it sounds like a song sparrow. They have an innate idea of what the song should sound like but they need the environmental feedback of hearing themselves sing – an 'innate schoolmarm' as Lorenz called it.

So, the intuitively reasonable distinction between 'innate'

and 'learned' based on 'sources of information' is difficult to maintain, at least if we insist that it must be hard and fast. Despite this, the word innate or at least the concept refuses to disappear altogether. It still appears in the literature, sometimes modestly clothed with quotation marks ('innate'), sometimes awkwardly as 'what we used to call innate'. People grope towards a category of behaviour, even though it may be a difficult one to define precisely because they want to emphasize one of the most remarkable features about animal behaviour – the fact that an animal may behave in a manner appropriate to its survival and reproduction the first time it finds itself in a particular situation.

A hard-line attitude would be to say that although that is an important point to make about animal behaviour, making it is hindered rather than helped by the label 'innate'.

But whether or not we use the label, there clearly is some behaviour that does have special properties. This behaviour develops in much the same way whatever the animal's early environment is like. No special experience is necessary for the animal to show appropriate, adaptive behaviour. The singing of *Teleogryllus* crickets is an example. A great deal of confusion has arisen, however, from the common belief that behaviour showing these properties of being unlearnt, adaptive, and 'developmentally fixed' must also, logically, be 'genetic'. Our next task is to show why this is not so.

'Developmentally fixed' does not mean 'genetic'

We have already seen that the development of behaviour in the individual should be distinguished from differences between individuals making up a group or species. Development can be followed in single animals. We can watch them grow up. We can try to isolate the factors that influence their development. But genetic differences between animals ('genes for' different characters) cannot be studied in single animals. There must be more than one for the idea of 'differences' to make sense at all.

This has an interesting consequence. It means that heavily selected traits – those for which natural selection has favoured only one genotype – tend not to exhibit genetic differences because only one successful variant may now remain. All individuals may be genetically similar for those traits because any individuals that were genetically different will have been selected against. Natural selection could have deleted them, leaving a genetically homogeneous population with, by definition, no 'genes for' those characters. As we have seen, this is not the same as saying that those characters are not influenced by genes. It simply says that there are now no differences between individuals that can be ascribed to differences in their genotypes.

Many of the traits described by ethologists as 'innate' fall into this category. Most or all members of the species behave in a certain way. They all remove eggshells or sing a particular sort of song, despite considerable differences in what they experience when they are young. The behaviour is 'developmentally fixed'. But, if there is little or no genetic variation, it would not be possible to say that there are 'genes for' the behaviour.

In the case of singing in male *Teleogryllus*, we have seen that this could be said to be 'innate' in the sense of being environmentally fixed in the face of many environmental manipulations. All crickets of the species sing in the same way. Measured within one species, therefore, there is no genetic variation in singing. Just looking at one species, we would find no 'genes for' singing behaviour.

If instead of considering just one species, however, we look at differences in singing behaviour *between* species, then we can identify 'genes for' singing. There are genetic differences in singing behaviour between *Teleogryllus oceanicus* and *T.commodus* as we saw in the last chapter. Within one species, in other words, we can show that there is developmental fixity but the lack of genetic variation makes it impossible to identify any 'genes for' the behaviour. Only when we look at more than one species is it possible to find large genetic differences and conclude that there are 'genes for' the behaviour.

The apparent paradox between developmental fixity and genetics should now be beginning to resolve itself. Genetic differences (between individuals) are quite separate from developmental fixity (within an individual). To use the single word 'innate' to cover both or to assume that 'innate' can be equated with genetic, is a sure recipe for confusion, one that Lorenz's early papers unfortunately did a lot to create.

To find out whether the behaviour is developmentally fixed, we have to study the effects of different environmental factors on development of individuals. To find out whether there are genetic differences, we have to do breeding experiments or resort to other ways of finding out whether similarities of behaviour result from similarities of genotype. Even then, as we have seen, the answer will depend upon the range of genotypes we have chosen to study. The secret is to demand always two sets of facts – one about genetics and the other about development. If we do this, we will not go too far wrong. But if we are content with just one set, such as evidence that the trait is heritable under some circumstances and believe that this has told us anything about modifiability during development, we will be led, very surely, astray.

As if this were not enough confusion for a single word to generate, there is yet another meaning to the word 'innate', arising from its association with 'instinct'.

Innate behaviour and instinct

The word instinct means literally 'driven from within'. In other words, it refers to the inner drive or motivational force that leads an animal or person to behave in a certain way. Lorenz was very careful to point out that in discussing innate behaviour, he was *not* talking about 'instinct' as such, but about the *'Instinkthandlungen'* – the behaviour patterns that resulted from the inner drive. He positively dissociated himself at first from the earlier controversies about drive and instinct that were to be found in the works of Freud, McDougall, and others. Later, however, he noticed a striking feature of almost all the instinctive behaviour patterns he studied: if an animal had not performed a particular behaviour for some time, it showed a greater tendency to do so when the opportunity next arose; if it had not eaten, it would show a greater tendency to eat when given food; if it had not courted a female, it would show a greater tendency for sexual behaviour, and so on. It was, to quote his own words, 'as though some response-specific energy were *accumulated* during periods when a specific pattern is not performed' (1937).

He thus began to see instinctive behaviour patterns as

driven by something inside the animal, something which built up and was discharged again when the animal performed the behaviour. If the animal could not perform the behaviour, it might search for the opportunity to do it (it might look for food or a mate) and if the right opportunity were still not forthcoming, the animal might make do with inadequate substitutes. It might try to mate with its food dish if no female were to be found. The internal motivation to do the behaviour was always there, building up and discharging and generally governing the animal's behaviour. Lorenz began to see this as yet another characteristic of instinctive behaviour patterns.

Now it is important to emphasize that the question of whether a behaviour is internally driven in the way Lorenz proposed is quite separate from the question of whether it is 'innate' or learned. One is a question about immediate causation – what makes the animal behave in this way or that. The other is a question about development – what factors, perhaps acting over a long period of time – made the animal the way it is. It is like the distinction between asking how a car works and how it is made. The difference is that in the case of cars, it is quite easy to keep the questions separate. In the case of animal behaviour, they often become inextricably tangled up, but it is very important to try and keep them separate.

For example, an animal may show 'innate' (meaning unlearned, developmentally fixed) anti-predator behaviour. It may respond quite appropriately with complex avoiding behaviour the first time it is exposed to a predator-like stimulus. And yet this would not imply that its anti-predator behaviour was necessarily driven from within and that the animal would search out predators to run away from if it had been deprived of the opportunity for doing so for some time. On the whole, anti-predator does not exhibit the 'driven from within' properties implied by the literal meaning of instinct. External stimuli (especially the presence of the predator) tend to be very much more important. Yet, many animals have the ability to avoid predators without having to learn to do so and whatever environment they are reared in. The 'instinctive' (driven from within) and the innate (unlearned, developmentally fixed) aspects of behaviour should therefore be thought of as quite distinct.

Should we talk about 'innate' behaviour at all?

It is small wonder that the word 'innate' has been the object of so much argument and debate. Two people arguing about it might each be meaning quite different things. One, thinking that innate meant 'entirely due to genes', might criticize his opponent for subscribing to a vacuous sort of genetic determinism. The other, meaning only that there are some things all animals of a species do without any opportunity to learn, might be rather taken aback at the vehemence of the attack on him. It would take a very long time to explain to each of them where their differences originated.

One possible solution would be to suggest that neither of them should use the word 'innate' at all and insist that they should explain, without once using that term, exactly what they each really meant to say. In other words, because the word innate has been responsible for creating so much confusion, there is a good case for abandoning it altogether. But it would be a very retrograde step to abolish the word and then fall back on the bland statement that all behaviour is a complicated interaction of genetic and environmental factors (a complex nexus as it has been called). This gets us nowhere in understanding how behaviour develops.

It also obscures a very important fact. Animals do many things which they can have had no opportunity to learn how to do. They are born or hatch out of an egg and immediately behave in ways that help them to survive. Their behaviour, which makes them seek shelter or find food, is as much part of their equipment for survival and as much a product of natural selection as any scale or spot or feather. This we should not forget, even though 'innate', with its motley retinue of meanings, may be the wrong word to be trusted with such an important message.

Further reading

Dewsbury (1978, ch. 7–9) and Bateson (1983a) give helpful accounts of the difficulties with innate behaviour. Hailman (1969) shows how learning can interact with innate behaviour.

6
Some obstinate remnants:
Instinct, fixed action patterns
and displacement activities

Capuchin. Photograph by Fritz Vollrath.

In the last chapter, our concern was to chart the difficult
connection between genes and behaviour, and we touched only
in passing on the idea of instinct in its literal sense of 'driven
from within'. Most contemporary textbooks on animal behaviour tend to dismiss 'instinct' altogether and attempt to con-
sign it to honourable retirement, together with 'fixed action

patterns', 'vacuum' and 'displacement activities', 'releasers', and several other derived concepts once in common use.

The trouble with 'instinct' and these other terms it has given rise to is that they will not go away. They still intrude themselves into the way we talk about animal behaviour, often without our realizing what implications they are bringing with them. They certainly form a major part of what outsiders think the study of behaviour is about. But, more seriously, they have recently assumed a kind of respectability in the scientific literature on applied ethology (e.g. van Putten and Dammers 1976). Lorenz's ideas of instinctive energy damming up inside an animal if it cannot perform a certain behaviour, for example, has been used to argue that hens suffer if kept in battery cages where there are many sorts of behaviour they cannot perform. Hens going through the motions of dustbathing on the bare wire floors of the cages have been said to be showing 'vacuum behaviour' (Vestergaard 1980).

So, because of their continuing use to the present day, and the confusion they still cause, we will spend this chapter looking at some concepts that might otherwise be thought to be of only historical interest. Seeing precisely what the objections to them are will also pave the way for our discussion of mechanisms in the next chapter. We will start with 'instinct' itself, and then move on to two terms closely associated with it and most widely in use today – 'fixed action patterns' and 'displacement activities'.

Instinct

Classical ethology, under the influence of Konrad Lorenz and Niko Tinbergen, saw animals as having forces inside them – 'instincts' that propel or 'drive them from within'. Lorenz's (1954) view was that 'the nervous system spontaneously produces energy parcels which are allotted to certain highly specific motor patterns, on the one hand developing a general drive (appetite) for their elicitation and on the other hand lowering the stimulus threshold for such elicitations' (p. 205).

Tinbergen (1951), more tentatively, wrote about 'the internal factors which determine the "motivation" of an animal, the activation of its instincts' (p. 57). Yet for Tinbergen, too, the nervous system of an animal spontaneously generated internal impulses that were an important factor in the production of coordinated behaviour. Inside the animal were forces that activated or drove it to behave in certain ways.

At the time, the idea was quite revolutionary. The prevailing opinion in physiology was that animal behaviour should be seen as a series of reflexes, with the animal reacting to a series of external stimuli. Coordinated behaviour was held to be the result of a chain of these reflexes in which performance of the first small step in the movement would stimulate the sense organs to activate muscles to produce the next step, and so on. The animal was thus seen to be essentially *re*acting to stimuli in its environment, rather than being spontaneously active of its own accord.

Here was Lorenz putting the emphasis quite the other way around. He saw the animal's nervous system as the active party and instinct as energizing it from within. He did acknowledge some role for stimuli outside the animal, because it is quite obvious that the things around them do have an effect on what animals do. But it was the build up of some instinctive factor or nervous energy inside the animal that he saw as the driving force behind the behaviour. The longer an animal had been without doing a particular kind of behaviour, such as feeding or aggression, the higher the levels of this nervous energy were supposed to become. The levels could build up to such an extent that the animal might start performing the behaviour without the usual external stimuli being present. It might start trying to mate with its food dish, for instance. There might even come a point when the instinctive nervous energy became so pent-up that the animal did the behaviour 'in a vacuum', apparently without prompting from any external stimuli. Lorenz referred to this kind of behaviour as 'vacuum activity'. He held that each different kind of behaviour had energy specific to it. Exactly how many instincts there were was left open, but they corresponded to whole categories of behaviour such as eating, fighting, and sleep (Tinbergen 1951) rather than more specific ones, such as different kinds of fighting behaviour.

The appeal of Lorenz's proposal was that it attempted to give a single explanation for very different kinds of behaviour – in

broad outline at least. Obviously the causal mechanisms for
sleep and fighting would be different in detail. There would be
different hormones released, different nervous mechanisms at
work and so on, but all these diverse mechanisms were seen as
having important things in common. They were all held to
share the common characteristics of being driven from within
the animal, with energy being built up and released when the
animal encountered the right external stimuli. Instincts,
generated from within, underlay everything.

Criticisms of instinct

Lorenz formalized his ideas about instinctive nervous energy
in a model that said that the behaviour of animals had some of
the same properties as a cistern being filled with water and
emptied through a series of valves (Fig. 3). The effect of time,
where an animal becomes more likely to perform behaviour
the longer it has gone without doing so, was mimicked by
water building up in the cistern and reaching a higher level
the longer it had been flowing in. The weight of a high level of
water could push out the valves at the bottom of the cistern
without anything pulling on the valves, just as a frustrated
animal might perform a vacuum activity without the need for
external stimulation, and so on.

Unfortunately, all instinct theories, at least those which see
instinct as a force driving the animal from within, suffer from
the same fatal flaw. It is not just Lorenz's idea of instinct as
'energy' or his analogy with cisterns and water that are at
fault, but any concept of an energizing force inside the animal,
whether it is called 'instinct', 'drive', 'action-specific energy', or
what. All of them suggest pent-up physical energy, like the
exploding gases in the cylinders of an internal combustion
engine, where the energy actually powers the engine.

But instinctive drive or Lorenz's action-specific energy does
not power muscles. Chemical energy does that. So 'instinctive
energy', if it means anything, must refer to something else,
perhaps the sum total of internal causal factors (hormones,
neural responsiveness, etc.) that exist at any one time. We
could talk about the animal having a high level or a low level
of these factors, but even a high level would not mean that the

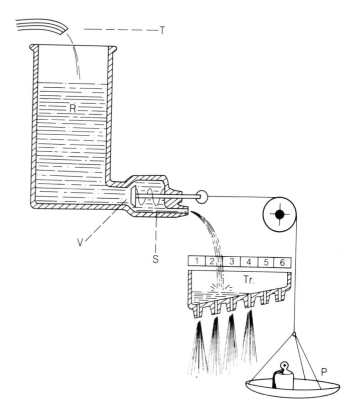

Fig. 3 Lorenz's psychohydraulic model of behaviour. Action-specific energy (represented by water) builds up in the reservoir R through the tap T as a result of deprivation. No behaviour is expressed while the valve V is kept shut by the spring S, but when the spring is moved, the valve is opened, the water rushes out, and behaviour occurs. The spring is moved by a combination of the weight of water in the reservoir (the strength of internal drive) and the weights in the pan P (which represent the strength of external stimuli). Different kinds of behaviour are determined by a perforated trough (Tr). If V is only partially opened and only a small flow of water is allowed out of the reservoir, only a small amount of water enters the trough and only the behaviour with the lowest threshold (1) occurs. If larger amounts of water leave the reservoir, the trough will become fuller and higher threshold actions (2–6) will occur. (From Lorenz 1950)

animal was necessarily going to be energetic or rush round in a frenzy. To talk about these causal factors as 'energy' or to say that the animal is impelled by instinctive drives implies that frenetic activity will result.

Often nothing could be further from the truth. A high level of causal factors for sleep results in inactivity. A sleep 'drive' 'energizing' the animal becomes a contradiction in terms and shows up the error of careless thinking about energy concepts. Animals do not have *energy* accumulating inside them. Various changes may take place inside their bodies and these changes may get progressively larger as time elapses. But they are not like a piston in a cylinder propelled to move by the sheer force of gases impinging upon it. To think of animals like this is to make a serious error about the nature of the very complex processes that are going on inside their bodies. If we want to understand these processes, it is better not to be misled into thinking that instinct 'propels' behaviour, attractive and superficially plausible though that idea may at first seem.

Hinde (1960) led the attack on 'drive' and 'energy'. 'Instinct' suffered from all the ills that he revealed in 'drive' plus another of its own: its implications of innateness. The last chapter dealt with the controversies that have surrounded that idea. Instinctive behaviour, with its 'inborn' as well as its 'driven from within' elements under attack, should not have survived at all.

But it has not disappeared entirely, however much its critics would like to see it go. It has such a powerful appeal that many people outside ethology still refer to instinctive behaviour in animals and are under the impression that that is what ethologists study. More interestingly, some other concepts that are still in current use owe their existence to the assumption that there are instincts and, more than that, that they have at least some of the properties proposed by Konrad Lorenz in his cistern model of instinctive behaviour. For the rest of this chapter we will discuss some of those ideas, the legacy of instinct.

Fixed action patterns

According to Konrad Lorenz's cistern model of animal behaviour, the end result of the release of instinctive energy is the performance of a sequence of behaviour. Konrad Lorenz and Niko Tinbergen both believed that the behaviour of all animals is made up of combinations of these sequences, which they referred to as 'fixed action patterns', a term still used today. Fixed action patterns were held to have certain very distinctive properties. First, they were held to be fixed or stereotyped, that is performed in exactly the same way each time an animal did them and in the same way by different individuals of the same species. The courtship patterns of ducks are a very good example. The males display very distinctively in front of the females and they are so stereotyped in what they do that they look like clockwork toys, rising up in the water, shaking their heads, and dipping their bills – always in the same way for a given species.

Secondly, Lorenz believed that these fixed sequences were controlled in a particular way – namely, that they were 'driven from within' by the instinctive energy that was being released. We have already mentioned the fact that Lorenz's ideas about internal impulses in animals went against the mainstream views of physiologists in the 1930s, which was that sequences of behaviour arose from chains of reflexes. The view prevailing at that time was that if an animal was seen to be doing a series of movements, the most likely explanation was that the first movement in the series caused some change either in the outside world or in the sense organs of the animal itself, which then acted as a trigger for the second movement, and so on. Lorenz proposed, in contrast to this view, that the whole sequence came from within the animal, without the need for external prodding or stimulation by the sense organs at each stage. The whole fixed action pattern was thus generated from within the nervous system. This mechanism became a more important distinguishing characteristic of a fixed action pattern than the fixity itself, even though that 'fixed' is part of its name and 'generated from within' is not.

At the time when Lorenz first discussed the causal basis of fixed action patterns, the physiological mechanisms under-lying them were largely unknown. Lorenz was struck by the stereotypy of the behaviour he was observing and simply hypothesized about mechanism. But more recently, in-vestigations into the workings of the nervous system – particularly in invertebrates – suggest that nervous systems may have at least some of the properties that Lorenz suggested. There may not be instinctive energy inside the animal, but instructions to the muscles about when and how far to contract may be generated within the nervous system itself.

There is a sea slug, *Tritonia*, which has a very stereotyped way of escaping from its predators, which are mostly starfish. It flaps its body up and down in a way which, although it is a bit clumsy, is usually enough to get it to safety (Fig. 4). The flapping lasts for about 30 s and is always performed in exactly the same, fixed way by the animal flexing first the dorsal and then the ventral muscles of its body wall.

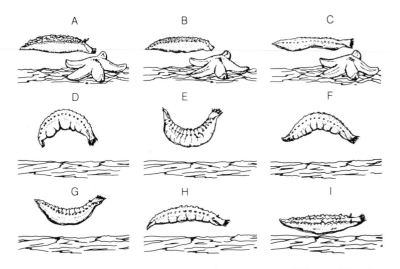

Fig. 4 Escape swimming response of the sea slug *Tritonia*. Chemical substances from a starfish (1) cause *Tritonia* to pull in the extended parts of its body (2) and then to elongate its head and tail regions into paddle-like structures (3). It then begins a series of ventral (4) and dorsal (5) flexions, giving rise to vigorous swimming movements. The response ends with a series of gradually weakened upward bends.

Now, this response could theoretically be brought about by a series of reflex movements. Perhaps the movement of the body into an upward curve could stimulate stretch receptors which then trigger the flexing of the opposite muscles into a downward curve, so that the animal goes from 'Ω' to 'U' to 'Ω' again because the result of the first movement is to stimulate the animal to do the second, and so on. This possibility can be excluded, however, by recording the messages of the nervous system when it is prevented from receiving information about the results of earlier stages in the movement (Willows, Dorsett and Hoyle 1973).

When the sea slug is showing 'escape swimming' as the flapping up and down is called, its brain sends out a series of messages to the muscles of the body wall, telling first one group to contract, then another group. These messages are quite distinct and easily recognizable. Then, if the brain of a sea slug is removed and put in a dish by itself, it is possible to make it send out the 'escape swimming' message by stimulating one of the nerve stumps still attached to it (Dorsett, Willows and Hoyle (1973). The rest of the brain's 'escape swimming' messages are almost exactly the same as in the whole animal during a real escape. The messages go out as if to the muscles, only now there are no muscles to receive the messages.

Feedback from the muscles or their stretch receptors in an earlier part of the sequence is thus clearly unnecessary for the remainder of the instructions to be issued. It is of course possible that in the intact animal the brain does respond to things that are happening in the muscles or elsewhere, but it does not have to. The nervous system itself generates a long series of instructions to the muscles 'from within'.

There are many other examples from both vertebrates and invertebrates which now suggest much the same thing, namely, the internal activity of the nervous system (Bentley and Konishi 1978). Feeding in snails, singing in crickets, and swimming in fish all justify Lorenz's claim that the patterns for many different behavioural sequences arise from within the nervous system, without the need for constant feedback from the environment. There may not be energy building up and being discharged but from within the animal comes patterned activity – instructions to the muscles about when to contract. The instructions are carried out like a clockwork toy going through its motions once the mechanism has been started off.

All this might lead us to think that fixed action patterns are a reality, in tune with the latest advances in neurobiology. In one sense, they are. The term 'fixed action pattern' is indeed sometimes to be found in the literature on invertebrate neurophysiology applied to such cases as the sea slug swimming. But for ethologists there are three very good reasons for steering clear of the term while acknowledging that Lorenz may have been right to emphasize the very positive role of the nervous system in giving rise to patterned behaviour.

The first of these reasons is that although we now know something about the physiological basis of some behaviour patterns and are in some cases able to point to a very active role for the nervous system in giving rise to them, it does not follow that all behaviour patterns arise 'from within'. Many behavioural studies have been of the complex behaviour of fishes, birds, and mammals where virtually nothing is known of the physiological bases of the behaviour. We know little about the physiology of courtship in sticklebacks, eggshell removal in gulls, or fighting in red deer stags.

Yet there are many questions to ask about such behaviour: why does the male stickleback court this female but chase away the next one?; what does a gull gain from removing eggshells from a nest?, and so on. We therefore need a term to describe the behaviour we observe that does not commit us to a particular view of its mechanism that the term 'fixed action pattern' carries with it. Calling something a fixed action pattern does not, as we have just seen, mean that it is fixed or stereotyped. It means that the pattern of instructions to the muscles is generated from within the animal's own nervous system. Because in most cases we do not know whether this is true or not, it is best to steer clear of a term that prejudges the issue.

A second reason for not using the term fixed action pattern, is that, even in those cases where the nervous system is very active in producing behaviour, there often turn out to be large effects of the environment as well. When water snails feed, the twenty-five pairs of muscles that control the mouthparts are ordered by messages from within two knots in the nervous system near the front end of the animal known as the buccal ganglia (Kater and Rowell 1973). As in the case of the escaping sea slug, the knots of nervous tissue can be completely removed from the animal and still go on issuing the same messages as they do in an intact animal. In this sense, feeding movements in snails are another example of a fixed action pattern.

However, if the snail tries to dislodge a particularly hard bit of food with its mouth, extra power is applied by a slight alteration in the way in which the muscles of the mouth contract. The nervous system is thus responsible for generating the basic message to the muscles from within itself, but environmental input is also incorporated during the sequence as the need arises. Such a situation is very common. It may be quite difficult to make a clear-cut distinction between patterns of nervous activity generated from within the nervous system but with some help from the environment on the one hand, and a chain of reflexes on the other.

Most animal behaviour simply refuses to fall into neat categories, such as fixed action patterns generated 'from within' or chains of reflexes patterned from the environment. On closer analysis, many so-called fixed action patterns turn out not to be so fixed after all. What looks like a rigidly stereotyped behaviour may in reality be highly modifiable. The drinking behaviour of chicks and chickens looks very fixed – the drinking takes place over and over again as if the animal were running on clockwork. But film analysis shows that there is considerable variability from drink to drink (Dawkins and Dawkins 1973). In fact, most animal behaviour turns out to be very variable. Dogs barking, snakes striking at prey, lizards displaying to rivals all show this variability (Barlow 1968; Stamps and Barlow 1973). 'Fixity' – the absolutely identical performance of a behaviour pattern over and over again – turns out to be the exception rather than the rule. It is certainly not a universal feature of animal behaviour.

The third objection to using the term fixed action pattern arises from its association with instinctive models of behaviour. The fixed action pattern was supposed to be the end result of the release of action specific energy. It was supposed to be innate or inborn and characteristic of all members of a species. Fixed action patterns are not only associated with Lorenz's model of instinct, they are part and parcel of it. We have already seen why this model and indeed all instinct models are not generally favoured in the 1980s. Fixed action patterns have come to be regarded with a similar scepticism.

'Fixed action pattern' is a term that is rarely used today by ethologists (although physiologists do sometimes use it). Barlow (1968) suggested that we should talk about Modal Action Patterns to get away from the idea of rigid stereotypy (which we often do not see) and from any implication of underlying mechanism (which we often do not know about). But most people content themselves with talking about 'action patterns' or 'behaviour patterns'. These are neutral terms, nothing more than convenient labels for observed sequences of behaviour. They carry none of the overtones of 'innateness' or undercurrents of 'instinct' that 'fixed action patterns' can never get away from.

Displacement activities

One of the most widely known offshoots of the Lorenz–Tinbergen theory of instinct is a class of behaviour called 'displacement activities'. The man in a quandary scratching his head is the one everyday human example which would be most likely to spring to mind. A classic bird example are the black-headed gulls observed by Tinbergen (1952), which, in the middle of a fight, momentarily stopped being aggressive to one another and started pulling up grass instead. 'Grass pulling' appears to have nothing to do with fighting – in fact, it looks more like nest-building – and it is thus somewhat odd that it appears at a time when one might have expected the attention of the two contestants to be engaged solely in fighting.

It was this 'oddity' of displacement activities that seemed to the early ethologists to need an explanation. A male stickleback, caught between attacking a female as an intruder into his territory and courting her as a mate, will often spend quite a bit of time doing neither, but will visit his nest and 'creep through' it, leaving the female at the edge of his territory where, one would have thought, she would be in danger of going away altogether (Wilz 1970). Male junglefowl, like gulls, also behave 'oddly' in fights. They will start pecking the ground and picking up small particles with meticulous care at a time when one would have thought that this was the last way to win a fight or keep an eye open for an opponent that was about to attack (Kruijt 1964).

Many textbooks define displacement activities by this oddity or 'irrelevance', that is its unexpected appearance when the animal appears to be engaged in something else altogether. But this is to ignore the true origin of the term. Tinbergen (1951) and Kortlandt (1940) independently proposed the same explanation of this 'odd' behaviour: nervous energy was 'displaced' from one instinct to another. The energy that had accumulated for one kind of behaviour and was unable to find an outlet leapt across or 'sparked over' and started powering a completely different sort of behaviour that the animal was able to do.

This situation was envisaged to occur if an animal had, say, a high level of energy driving it to attack a rival but an equally high level of energy in another system (cistern) driving it to run away. The animal would then be highly motivated to do two kinds of behaviour at once, but two kinds of behaviour that were, in this case, completely incompatible. Nervous energy was thought to leap or 'spark over' from the system in which it was building up (attack and escape in this example), into a third system, say, that controlling nesting behaviour. So if a gull started showing elements of reproductive behaviour – trying to gather grass to build its nest – this 'grass pulling' would be explained as being driven or powered by the dammed-up energy from the aggressive and retreating behaviour. The junglefowl cock, also stimulated to retreat and attack simultaneously, would similarly be seen as having energy from these behaviour patterns channelled into the feeding system; on this occasion, feeding would appear driven by energy 'displaced' from elsewhere.

To label an action as a 'displacement activity' is, then, to state a commitment as to how it is caused – namely, by the displacement of energy from one motivational system to another.

'Displacement activity' is a term that has passed firmly into our language, so much so that we often do not realize what its implications are. It is not just a description of the behaviour nor even of its 'oddity' or 'irrelevance' but declares its dependence on the Lorenz model of instinct loudly and clearly whenever it is used. It should have been abandoned when the drive and energy models themselves were found to be so seriously lacking as explanations of behaviour, but it has lingered on. Even if we try to separate it from its motivational implications and concentrate on irrelevance or oddity of certain kinds of behaviour, there are problems.

Behaviour that at first sight looks odd or irrelevant to us, as naive observers of what the animal is doing, may turn out to have very great relevance. For example, junglefowl cocks that 'oddly' peck at the ground during a fight actually seem to be the ones most likely to win the fight in the end. If denied objects to peck at, they are not so likely to win (Feekes 1972). And male sticklebacks that take 'time off' to go to their nests while keeping the female waiting at the edge of the territory also seem to do better in the long run. They either court more successfully when they do return to the female (Wilz 1970) or have a better nest (Cohen and McFarland 1979).

In other words, we should not jump to the conclusion that the behaviour we are watching is 'irrelevant' just because we cannot immediately see what the relevance is. Animals are always doing things that surprise us in the sense that however long we have spent watching them we can rarely predict exactly what they are going to do next. 'Displacement activities' are thus no more mysterious than much of the rest of what we see animals doing. But the term itself is a distinct nuisance. It is not just an interesting etymological oddity like saying 'good-bye', which probably originated from 'God be with ye', when we take leave of people. The religious origins of this phrase do not generally trouble us. But 'displacement activities', with their blatant message of instinctive energy being displaced from one system to another, are not so harmless. They crop up, for instance, in discussions of animal welfare with the clear implication that if a captive animal shows displacement activity, that is a sign that there is something wrong with the way it is being treated. If energy is being 'dammed up' and the animal shows displacement activity, then, it is argued, it must be suffering through being unable to express other behaviour (van Putten and Dammers 1976). Such oversimplistic thinking about the way in which animal behaviour is motivated arises directly from a continued belief in Lorenz's model of instinctive behaviour in general and from the idea that energy can be dammed up and 'displaced' in particular.

Conclusion

One thing that will be apparent from this chapter is the sheer magnificence of Lorenz's model of instinctive behaviour. It may be wrong in certain respects, but it has a grandeur in its attempt to give an explanation for all animal behaviour that we do not find in most theories of smaller scope. Lorenz and Tinbergen divided all animal behaviour into fixed action patterns. These were powered by energy that accumulated like water in a cistern. Their model attempts to account for periods of exhaustion after an animal has performed a behaviour as well as for the increased tendency that the animal often shows when it has been deprived of the opportunity for performing it for some time. The model also has an explanation for behaviour that occurs in the absence of the right external stimuli and for 'odd' behaviour that animals show when they are in a motivational conflict or prevented from doing something. It sees much behaviour as inborn, stereotyped, and characteristic of the species. Both Lorenz and Tinbergen emphasized that animal behaviour is as much shaped by the action of natural selection as any morphological character, and that its evolution could be studied. Function, causation, and development were united in their single conceptual framework.

We cannot help but respect a theory that attempts to explain so much and tries to account for so many aspects of the behaviour of so many animals. A theory that explains the Universe inspires more admiration than one that explains just a small corner of it. And the Lorenz–Tinbergen model is no small-corner explanation. It has inspired a great deal of thought and research even though, perhaps precisely because, it has always been controversial.

Nevertheless, the criticisms that we have discussed and that have led to its being replaced as an explanation of behaviour are substantial and we should not allow ourselves to indulge in careless use of such terms as instinct, fixed action patterns and displacement activities which are so directly derived from it.

Further reading

Tinbergen's (1951) *Study of Instinct* describes the classical ethological approach. Hinde (1970, ch. 17) gives a very useful account of displacement activities as well as other categories of behaviour. Barlow (1968) discusses fixed action patterns from an ethological point of view, and Bentley and Konishi (1978) provide a neurophysiological account of pattern generation.

7
The machinery of behaviour

Silver argiope spider (*Argiope argentata*) wrapping prey. Photograph by
Fritz Vollrath.

Lorenz's view of instinctive energy inside an animal driving it
to perform behaviour may no longer be generally accepted, but
people still try to work out what the internal 'machinery of
behaviour' might be. In doing so, they have, once again, fuelled
controversy and given rise to confusion.

The controversy is now not so much about the value of a
particular model but how to go about trying to understand the
workings of the immensely complex machinery of behaviour.
One view – that of neurophysiologists – is that the best way to
understand it is to look inside the animal's body and try to see

how the various cells and organs function. The other view – the 'black-box' or 'whole-animal' view – is that knowing how individual parts work is not and never will be sufficient to understand behaviour and that what is needed is to study the behaviour of whole, intact animals without opening them up.

At times, this controversy becomes quite acrimonious, with each side claiming that their approach is not only the best but actually the only sensible way to study animal behaviour. To see our way through this controversy, we will start this chapter by looking at what these various approaches are, beginning with the ambitious attempts to map the connections of individual nerve cells that have been carried out on invertebrates and ending up with the whole-animal or 'black-box' studies which have been done without the aid of any direct physiological measurements at all. Knowing what the approaches are will enable us to see more clearly the value and limitations of each. When we get to the whole-animal approach we will have to spend a certain amount of time sorting out some of the confusions that have arisen over terms such as 'motivation' and 'drive'. We begin with what is perhaps the most straightforward approach of all – looking at the elements of the nervous system, the nerve cells, and working out how they give rise to behaviour.

The direct approach: mapping the nervous system

A major breakthrough in the understanding of the way the nervous system gives rise to behaviour came with the discovery that single nerve cells, at least in invertebrate animals, could be identified and shown to have certain characteristic roles in the production of behaviour. This idea, first put forward by Wiersma in the early 1950s, did not receive general acceptance until fifteen or so years later when the use of distinctive coloured dye enabled the connections of a given nerve cell to be traced clearly for the first time. Single nerve cells, far from being anonymous entities in a crowd of other neurones, began to be seen as having an identity of their own. Not only was a nerve cell revealed as having certain specific connections to other cells, but a very similar cell, with similar connections, would be found in other individuals of the same species.

We are not surprised to find that all individual insects of a particular species have six legs and similarly shaped jaws, but it was something of a surprise to find that they all have many individual cells of their nervous system the same too. The nerve cells and their connections can be mapped like the circuit diagrams of a computer and the maps apply to all individuals of a species.

One particularly well-known map is the jumping circuit of the locust. A cell known as the lobular giant movement detector (LGMD), found at the base of each eye, receives connections from different parts of the eye and responds to movement in the locust's visual field. Locusts also have another large cell called the descending contralateral movement detector (DCMD), which is connected to the LGMD. The DCMD connects directly to other nerve cells and these are responsible for activating the muscles which extend the legs for a jump. At the same time, the DCMD also connects to another set of motoneurones that activate flexor muscles in the legs, but it inhibits them from contracting. These inhibitory connections to the flexor motoneurones prevent the locust from trying to extend and flex its legs at the same time, so that the excitatory connections can activate the legs for the jump without interference.

So the route by which visual stimuli impinging on the eye lead a locust to jump has been at least partially mapped out (Burrows and Rowell 1973; O'Shea and Williams 1974). In other words, the 'machinery' of jumping in the locust – how a stimulus in the outside world leads to the behaviour of jumping away – is now known. The actual circuits are more extensive than we have space to go into here, but the general idea should be apparent. It is possible to probe the workings of particular nerve cells, trace their connections with other nerve cells and to learn a great deal about how the behaviour is brought about. So far, about 150 nerve cells in locusts have been probed and mapped (Hoyle 1983) and the circuits for some of the behaviour are now understood in considerable detail.

This 'circuit-breaking' approach, at least for insects and molluscs, appears to have been enormously successful. And yet, even among the euphoria of these successes, there is a despondency about the sheer magnitude of the problem. Each nerve seems to be subject to so many influences from so many other cells that some physiologists are daunted by the task of constructing comprehensive circuit diagrams for all the behaviour which an insect shows. Even Hoyle (1983), optimistic champion of the 'identified neurone' approach, has to admit that the timescale for completing the task is rather long. He puts it at several hundreds of years!

The combined physiological and 'whole-animal' approach

If insects and molluscs, with nervous systems containing a million or so nerve cells, present physiologists with a daunting task, most people pale at the thought of attempting anything remotely comparable for vertebrates. With birds and mammals, we are dealing with nervous systems containing many millions of nerve cells. A circuit diagram is almost beyond comprehension if each of these units is mapped individually. Even if it were available, it would be impossible to follow. So compromises are made. A small part of the animal is studied, such as the way the eyes send visual information to the brain, or the way the ears receive sound. And, instead of trying to map out the function of each individual nerve cell even in this small part, whole groups of cells with similar functions are studied together. The most fruitful studies are those in which the physiological probing of the nervous system is combined with studies of the animal behaving naturally. To illustrate this, we will take an example that we have already used in another context: bat echolocation.

As we saw before, bats like *Myotis*, the little brown bat, emit ultrasonic pulses of sound (Fig. 5) which bounce off objects around them and come back to the bats as faint, slightly distorted echoes. The distortions in the echoes are caused partly by what happens to the sound as it travels outwards and back (absorption of sound energy by the atmosphere) and partly by the objects themselves. Some objects will absorb certain sound frequencies more than others and so the echoes which bounce back off them are deficient in certain frequencies.

Bats are able to interpret these distortions to tell them what sort of objects gave rise to them. A hairy moth will give a very

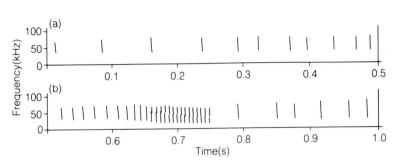

Fig. 5 Sound made by a little brown bat (*Myotis lucifugus*) while catching prey. Pulses of sound are produced while the bat is cruising around (a). As it homes in on the prey (b), more and more pulses are emitted, with shorter intervals between them, giving rise to the 'interception buzz' shown between the two stars. The bat then resumes normal cruising. (From Sales and Pye 1974)

different sort of echo to that of a shiny leaf surface, and a moving insect will have a different echo to a stationary one. This much bat echolocation shares with a human echolocation system (sonar) for detecting how far the ocean floor is from a ship. What is staggering about the bat's echolocation system is the rapidity with which pulses of sound follow one another and the fact that the bat is able to listen to the echoes in the millisecond intervals between its own noises.

Over a hundred times a second the bat shouts, listens, and shouts again. It is able to gather information about how far away objects are by using the time which the echo takes to come back to it. It can accurately localize where the object is and what it is (Simmons et al. 1979). Some bats, for example horseshoe bats, can even gauge how fast an object is moving and whether an insect is moving towards or away from them (Schnitzler 1973). Understanding the machinery of bat behaviour is clearly quite a task.

The initial understanding of how the mechanisms involved came from behavioural observations on what bats do naturally. Griffin (1958) looked at the behaviour of free-flying bats to establish that bats really can echolocate – that is find their way around using only the echoes of their own voices, without using vision, smell, or any other sense. Just recording the sounds that bats make when they are catching insects was highly instructive in itself.

A little brown bat cruising around and not apparently in pursuit of a particular prey emits pulses of sound about 50–80 kHz in frequency and makes about 25–50 of them in each second. Each pulse is very short – only 1–2 ms. Then, as the bat homes in for the kill, its sounds change. The pulses of sound become even shorter (down to 0.2 ms) and more and more of them are packed into each second. As the bat is about to strike its prey, it may be making 200 separate sounds in each second.

This observed change in the sound that a bat makes as it comes closer to its prey was what originally suggested a mechanism by which it might gain information about how far away the prey is.

As a bat closes in on an object, the echo will come back to it more and more quickly. So if the bat keeps emitting pulses of sound at the same rate and of the same length that it was when it was cruising around, then it would find itself in the position of still making a loud sound while the previous echo was coming back to it. Only if the sounds that it was making are shortened and timed to occur between the returning echoes could it hear the faint echoes at all.

Behavioural experiments confirm that bats do indeed use the time an echo takes to come back to them to tell them how far away objects are. Bats can be trained to fly to the nearer of two landing platforms if they get a mealworm for doing what is required of them (Fig. 6). Bats relying only on echolocation can learn to fly to the nearer platform 58.7 cm away even if it is only 13 mm nearer than the 'distant' platform at 60 cm (Simmons and Vernon 1971). Now, a difference of 13 mm would give rise to a difference in echo arrival time of only 75 µs. In other words, an echo bouncing off an object 58.7 cm away comes back to the bat only 70–75 µs earlier than an echo bouncing off an object 60 cm away. Yet this does seem to be what bats can do.

More recently, part of the physiological machinery for this behaviour has been discovered. Neurophysiological recordings from the auditory cortex of the big brown bat (*Eptesicus*) have revealed a class of nerve cells which can be described as

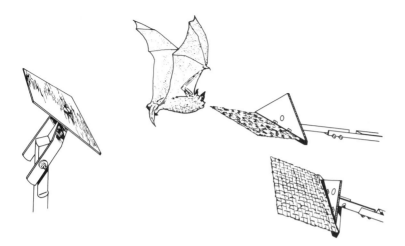

Fig. 6 Training a bat to discriminate between two targets. The bat flies from the starting platform on the left to whichever landing platform is associated with food reward. (From Simmons and Vernon 1971)

'echo-detectors' (Feng, Simmons and Kick 1978). These cells never respond if the bat is listening to just a loud, bat-like sound on its own. But they do respond if the bat hears a loud, bat-like sound and then, a little time later, a faint, bat-like echo. They do not respond to the echo alone unless it is preceded by a loud sound. They respond much less to loud 'echoes' than to softer ones – they prefer the sort of loudness a bat would normally hear when listening to a real echo. Some of the cells are very sensitive to how long a delay there is between the loud sounds and the following soft echoes. Some classes of cells respond most vigorously if this delay time is 4.5 ms, others if it is 20 ms. The difference between 4.5 ms and 20 ms corresponds to a difference in target distance of 0.75–3.5 m given the speed of sound in air. This means that there are different cells which are maximally responsive to echoes from objects at different distances. How far away an object is from the bat has thus been shown to be, at least partly, encoded in different places in its brain.

The discovery of 'echo-detecting' nerve cells illustrates both the strength and the limitations of attempting to tie down behaviour to particular cells or even classes of cells in vertebrates. On the one hand, it is a major step forward to have discovered that there are echo-detecting cells. On the other, we are almost as far as ever from being able to draw up a 'circuit diagram' of a bat. There are many things that bats do for which we still have no physiological explanation – their ability to adjust call duration and emission rate as they come into a target, for example.

But it is not just ignorance of physiology that has made people turn to other ways of studying the machinery of behaviour. It is not just the complexity of the circuit diagrams that would be involved in a complete physiological explanation. It is something much more fundamental, to do with the very nature of an 'explanation' of behaviour.

Whole animals and 'black boxes'

As we hinted at the beginning of the chapter, there is a genuine difference of opinion about the best way of understanding the mechanism of animal behaviour. There are those who, buoyed up by the success of circuit-breaking in invertebrate nervous systems, feel that their studies are paving the way for similar studies on vertebrates. The way to 'understand' is to look inside the animal, discover the physiology of the elements of its nervous system, and gradually build up a picture of how the parts interact.

But there are others who see the way forward quite differently. For them, circuit-breaking is all very well for simple behaviour and simple nervous circuits, but for most of the behaviour of interest to ethologists, it is not enough. A complete circuit diagram of a honeybee, even if we had such a thing, would not tell us that bees can communicate the location of food to their fellow workers. That we can learn only by studying the behaviour of whole animals (von Frisch 1967). And the mechanism by which the communication takes place – a 'waggle' dance in which both the direction of food and its distance from the hive – was discovered purely by looking at the behaviour of whole animals.

For many people, the way to study the mechanisms of complex behaviour in animals is not to look inside them as a physiologist would but to treat them as a whole and to try and work out what their internal mechanisms are from the way they behave. The animal is treated as what has come to be known as a 'black box', an initially mysterious object which is not to be opened but whose workings can be deduced from what it is capable of doing. This idea – of deducing what the mechanism is from the behaviour that results from it – is such an important one that we must spend some time going into it. It is impossible to understand why people have employed such concepts as 'motivation' and have deliberately tried to study mechanism without using any physiology at all and without first understanding what the value of this 'black-box' approach is.

To do this, we will use one of the most dramatic 'black-box' analyses of all time. Although it is an example drawn from genetics rather than behaviour, its results are so clear cut and revealing that it provides an object lesson for all studies of mechanism.

We tend to think that the greatest advances in genetics have come during the last thirty years from the discoveries of the structure of the genetic material itself – from a probing analysis of the elements involved rather than treating an organism as a 'black box'. But long before DNA was investigated or even known about, 'black-box' genetics had told us a quite extraordinary amount about genes, about chromosomes, crossing-over, and even the position of the genes along the chromosomes. The methods of 'black-box' genetics are simplicity itself: two organisms are mated and their resulting offspring counted.

From such unpromising results, Gregor Mendel was able to *deduce* that there were factors inside the germ cells of organisms that were responsible for its various characteristics. He was also able to deduce that these factors were quite discrete and distinct from one another even though they might temporarily coexist in the same body and interact in their effects. He also hypothesized that the factors assorted themselves into the next generation quite independently of one another. These are very powerful deductions. The important point for us is that Mendel hypothesized about genetic factors, segregation, dominance, and recessiveness by treating animals as black boxes and knowing nothing about the material basis of what was going on inside.

When, in the early part of this century, it began to look as though Mendel had been wrong about genetic factors assorting independently of one another, T. H. Morgan (1911) came to the rescue with another black-box explanation which told us even more about heredity.

Because characteristics of organisms were found to be 'coupled' more often than Mendel's theory of completely independent assortment should predict, Morgan suggested that genes went around in groups, later interpreted as being on the same chromosomes. And because the coupling is less than perfect, he further suggested that crossing-over between these chromosomal groups sometimes occurred. He went on to reason that the chance that two factors would go into separate germ cells when crossing-over would be a function of how far apart they were on a chromosome.

Sturtevant (1913) then argued that if Morgan were right, it should be possible to produce a map of genes along a chromosome using the chance of their being separated by crossing-over as the index of nearness.

So, long before anyone knew anything about how genes had their effects or even what the genetic material was, the position of the genes along the chromosome was known in considerable detail. Morgan was certainly aware of the descriptions of the movements of chromosomes that had been seen down the microscope and had in fact interpreted the finding that chromosomes twist around each other as the physical embodiment of the crossing-over he had postulated. But he was most insistent on the value of the genetic evidence in its own right: 'The proof of the linear order of the genes is derived directly from the linkage data', he wrote in 1919. 'It is not dependent on the chromosome theory of heredity.' And in describing genes, he wrote (p. 237), 'What I wish to emphasize is that their presence is directly deducible from the genetic results, quite independently of any further attributes or localization that we may assign to them.'

Genetics has come a long way since Morgan wrote those words. Many discoveries simply could not have been made just from deductions from genetic results. There comes a time in

the analysis of any system when it becomes necessary to discover the physical embodiment of the entities whose existence
has been deduced. But equally, there are many other things
which could only have been discovered through a black-box
approach. It was not molecular genetics that told us about
segregation and sex-linkage. It was black-box genetics. In the
same way, ethologists can still accept that, ultimately, all
behaviour depends on physiology, but that there is a great deal
we can learn only through 'black-box' ethology. What that
might be, we look at in the next section.

'Motivation' and mechanism

A bird feeds. Some time later it stops feeding and goes to sleep.
There is still plenty of food available and nothing that we can
see has changed in its environment. It appears to be something
in the animal that has changed. In view of the lessons of the
last few pages, it is almost certain that we will not have a
complete physiological understanding of what is going on for a
very long time. But some internal change must have taken
place to produce the overt change from feeding to sleeping. And
it is not just one specific kind of behaviour that is affected. The
animal has stopped pecking at food and it has also stopped
turning over leaves and actively walking around with its head
close to the ground. It is now doing a quite different set of
behaviour patterns – closing its eyes, fluffing its feathers, and
so on. Whatever it is that has changed has caused it to switch
from the set of actions associated with catching and ingesting
food to the quite different ones associated with sleep.

So, although for the moment we may be quite ignorant about
the physiological basis of the changes going on inside the
animal, we can nevertheless infer something about what these
changes must be like. They must, for instance, be reversible
because we observe that after a period of sleep, the animal goes
back to its previous behaviour of feeding. They must also be
changes that affect not just one behaviour pattern such as
pecking but a whole range of actions concerned with obtaining
food, because we observe that the animal changes from a
period of doing almost exclusively one set to doing another set
associated with sleep.

These changes – whatever they are – are referred to as changes in the animal's 'motivation'. Motivational changes are usually distinguished from those associated with injury or fatigue and also from longer term and more permanent changes due to learning or maturation. The bird in our example was undergoing motivational changes – short-term reversible alterations in its internal state. Now, 'motivation' is a deliberately vague term and it is also a declaration of ignorance. If we knew exactly how behaviour was produced, we would have no need to use the term 'motivation' at all. We would talk in terms of things like interactions between nerve cells and the release of hormones. But we do not have this knowledge and we are impatient. We do not wish to wait around for the several hundreds of years that physiologists are going to take to remove our ignorance.

Sometimes, fortunately, they do not take that long. On a few rare occasions, a motivational analysis of the behaviour of a whole animal can be followed up almost immediately by discovery of its physiological basis. As has happened before, a mollusc provides one of the best examples.

Pleurobranchaea is, like *Tritonia* that we met in the last chapter, a large sea slug. It is carnivorous and eats, among other things, other sea slugs of its own species. It eats almost whenever it gets the chance. It does show other behaviour as well, such as righting itself when turned upside down, mating, laying eggs, and withdrawing its head if its oral veil is touched. But feeding takes priority in a large number of cases and appears to inhibit a lot of these other activities as long as there is food available (Kovac and Davis 1977, 1980).

Normally, for instance, a *Pleurobranchaea* that is turned on its back will immediately right itself. If an upside-down *Pleurobranchaea* is offered food, however, it remains upside down and eats until the food is finished before turning the right way up. Without knowing anything about the physiological basis for this, we could say that the strong 'motivation' for feeding appears to inhibit that for turning the right way up. It is clear there is an interaction between feeding and righting but at this stage we know little more than that.

Feeding also appears to inhibit mating. If two sea slugs are offered food, while copulating, they break apart and start feeding. Withdrawal of the head is almost, but not quite as clearly inhibited by feeding. If the head of a sea slug is hit while the animal is eating, it will continue with its eating unless it is satiated or the food stimulus is not very strong. Then it does stop and withdraw its head.

Even feeding does not take absolute priority over everything, however. When the animal is laying eggs, feeding is inhibited, which makes functional sense because otherwise it might eat its own eggs. And escape inhibits everything. It is at the top of the sea slug's motivational priorities in the sense that if there is danger, all other activities are stopped and the animal attempts to get to safety. The adaptive significance of this is obvious.

The inhibitory connections were deduced purely by behavioural observations – treating the animal initially as a 'black-box'. Subsequently, the physiological basis for at least some of them were established. The inhibition of feeding by egg-laying, for example, is now known to be due to a hormone that affects the buccal ganglia (the knots of nervous tissue at the front of the animal that control the mouth movements). The way in which feeding inhibits withdrawal of the oral veil has been shown to be through a pair of neurones that are active when the animal is feeding. If these are stimulated, the usual messages to the oral veil that cause it to be pulled into the body are stopped (Kovac and Davis 1977).

With few animals do we have such a complete understanding of the interaction between different kinds of behaviour. Rarely can we go from black-box to physiological analysis so easily. But few animals appear to be so simply organized as a sea slug.

In many animals, we find that feeding takes priority when they have been deprived of food for some time but may well give way to drinking, fighting, or some other activity when the animal is only slightly less deprived. Fights between animals may be prolonged affairs with shifting priority given to attack and retreat depending upon each animal's varying assessment of its chances of winning (e.g. Maynard Smith and Riechert 1984). Simple rules like 'always eat unless attacked or laying eggs' are not sufficient to account for the behaviour of most animals faced as they are with a complex shifting world, with foods of changing quality, a territory under varying degrees of attack, a mate to be guarded, and unpredictable predators to be watched out for.

So it is not surprising to find that 'motivation' for different behaviour in different animals is very diverse or that black-box analyses of motivation have not given clear-cut results. Few animals are organized on the simple lines of a sea slug. Unfortunately, the response to this complexity has often been to postulate vaguely formulated motivational models which have had the effect of adding to, rather than removing, confusion.

The most serious confusions have arisen over the concept of motivation itself (Hinde 1959). Although 'motivation' was and always has been a vague term, there is often a tendency to believe that it is possible to talk about 'feeding motivation' or 'drinking motivation' as though they were two quite separate sets of mechanisms. This is clearly a mistake. When an animal eats, its water balance may be strongly affected, which in turn may influence when and how much it drinks. The two systems are clearly linked and feeding motivation cannot be considered in isolation from drinking motivation.

Another confusion arises because talking about a single 'feeding motivation' implies that many different behavioural actions can be placed under this one umbrella. 'Feeding', as we have seen in the feeding/sleeping bird example that started this section, is not just the act of taking food into the mouth, but in many animals also consists of a complex series of searching, stalking, and killing movements. Although these acts often occur in the same periods of time as each other, they by no means always do so. A cat may be so satiated with food that it will not eat, but it may still go through activities involved in hunting and killing (Leyhausen 1965). Eating and hunting thus appear to have rather different 'motivations'.

Similarly, when a rat is drinking, the amount of water it drinks, the number of times it will press a bar to obtain water, and the amount of quinine (which it finds distasteful) it will tolerate in its drinking water might all be thought to be measures of 'drinking' and all explained by the same 'drinking motivation'. But Miller (1957) showed that these different measures of drinking gave quite different pictures of 'drinking motivation' when rats are injected with salt. The amount of water drunk rises for 3 hours after such an injection and then levels off so that rats appear to be no more 'motivated' to drink after 6 hours than they are at 3 hours. On the other hand, the amount of quinine tolerated in the water steadily rises as time passes, so that the rat will drink a great deal more 6 hours after the injection than it will 3 hours after it. On this measure, it appears to be a great deal more strongly 'motivated' to drink after 6 hours than after 3 hours. It would clearly be quite misleading to talk about a simple change in 'drinking motivation' as if there were only one thing inside the animal that had changed.

'Motivation' is, therefore, a concept which is easy to invoke, particularly when we have no idea of underlying physiological mechanisms, but it is one that must be used with care if it is to tell us what is going on inside an animal. It is misleading to talk about a single motivational change if in fact several different factors are changing independently as oversimplistic views of 'drinking' or 'feeding' motivation clearly do. At the same time, black-box deductions about what goes on inside animals are an essential part of our understanding of behaviour. Knowing that 'feeding motivation' in a sea slug inhibited 'righting' and 'withdrawal' led to the search for the physiological mechanism by which this is achieved. Demonstrating, by behavioural means, the extreme accuracy of the bat's distance estimation mechanism, has shown what has to be explained in physiological terms. Without a knowledge of the behaviour and what an animal is capable of doing, the circuit diagrams of the nervous system, even if we had them, would be baffling in the extreme.

Conclusions

In a book of this length, it would be impossible for us to cover all that is known about the machinery of behaviour, because it is now such a vast and ever-growing subject.

We have seen that there is a considerable diversity of opinion as well as confusion. There are those who are convinced that the best way to move forwards is through physiology – finding the elements of the machine and trying to work out how they are connected. There are also those who think that a 'whole-animal' or 'black-box' approach is the only way to deal with the phenomenon of animal behaviour. They feel that the machinery is so complex that looking at the millions of parts that go to make it up will only confuse and make it more difficult to see how the whole works. They have resorted to such concepts as 'motivation', and not always clarified things in the process.

An unfortunate degree of chauvinism has crept into this debate. It often seems insufficient to say that one approach might contribute just as much as another. People sometimes seem to find it necessary to claim that one approach is the best or even the only worthwhile way of proceeding. Hoyle (1983), having described the mechanism for jumping in locusts that we discussed earlier, remarks triumphantly: 'These findings put an end to the 'black-box'/IdN [Identified Neuron] debate: the IdNers have prevailed! Already IdNers are setting the scene.'

Identified neurone-chauvinists are not, however, the only ones around. Oatley (1978) puts the opposite case just as strongly in an unashamedly biased (and very amusing) criticism of the neurophysiologists' method of finding how animal bodies work by looking inside. He argues that their methods are analogous to trying to find out how a computer works by 'cutting cable tracts, removing circuits boards and for larger scale operations removing substantial quantities of the computer with a shovel' (p. 25).

There is an important point here. Even a complete circuit diagram of a computer would not tell what it could do – that it could play chess or recognize shapes, for example. Understanding 'how' a computer plays chess involves understanding the program, the software that could in principle be run on computers of many different circuit diagrams. Similarly, understanding how an animal behaves is not just a question of knowing how it moves its limbs. We might understand how a bird moves it wings up and down but that would not have told us how it outmanoeuvres a hawk or finds its way back to the same nest site year after year with journeys of thousands of miles in between.

For, despite the information we now have about how the nervous system works, there is one thing we should not lose sight of. It is this: animals are far more complicated and versatile than any machine so far made by man. It would be a mistake to be too blinkered or too wedded to either the physiological or the 'whole-animal' approach when what we are trying to do is to understand the most complex phenomenon the earth has ever witnessed – the behaviour of animals.

Further reading

Ewert (1980) and Huber and Markl (1983) describe the neurophysiology of vertebrate and invertebrate behaviour, respectively. Hinde (1970 ch. 8–11) outlines the meaning and difficulties with some motivational terms. Bolles (1975) and McFarland and Houston (1981) give an account of more recent approaches.

8
Communication

Wandering albatrosses (*Diomeda exulans*). Photograph
by Michael Brooke.

Communication is the fabric of animal social life. It is the way
they influence one another to come together in schools, flocks,
and herds as well as to space out into territories. It is the way
the sexes interact in courtship, rivals settle disputes without
fighting, and often the way young animals obtain food from
their parents. In fact, looking at the way in which animals
spend their time, it is striking that many of them appear to
spend a great deal of it either influencing, or being influenced

by, the behaviour of other animals – in other words in some form of communication. It is not surprising, therefore, to find that people who study animal behaviour have spent a great deal of their time studying animal communication. What is surprising is to discover that, despite this intensive study, the whole subject is extremely confused, largely because of the definitions of various terms that have been adopted. In almost no other area of ethology have definitions so clouded the issues and so blinkered the research that has been carried out. Particular confusion has arisen over the definitions of 'signal', 'communication', and 'information transfer', and it is these that we will be looking at in this chapter.

Why 'communication' and 'signal' should be difficult to define

At first sight it is not at all obvious why there should be any difficulties with either of these terms. Both are words which are used frequently in ordinary language without any great misunderstandings appearing to result. But let us see what happens when they are applied to some of the most impressive examples of animals influencing each other's behaviour: flocking in birds and schooling in fish. In both flocks and schools, there can be such good coordination between the individual members that the group seems to take on a life of its own. The flock or school can wheel and twist about, all the individuals sticking closely together and all seeming to turn at just the same moment. Here, surely, is communication, even though we humans, looking in from the outside, may need slowed down film, or videotape to know what precisely the animals are responding to (Davis 1980; Potts 1984).

And yet, if we turn to almost any textbook on animal behaviour, we get a completely different picture. Animal signals, we read, have evolved so that they are conspicuous and exaggerated. They have become changed in the course of evolution so that they are more effective at altering the behaviour of other animals (Tinbergen 1952). Some signals clearly are like this. In the courtship of many bird species such as ducks, bright, conspicuous plumage in the male is combined with repeated stereotyped behaviour patterns to present the female with an eye-catching and unmistakable show.

But to insist that this is always the case when one animal influences another seems to fly in the face of the facts. Many animals clearly do influence each other by subtle, tiny movements so why do textbooks insist that conspicuousness and exaggeration are the chief characteristics of animal communication?

Most textbooks give a definition that appears to be relatively straightforward to apply to animals. The definition is usually based on the idea that communication occurs when one animal can be shown to have an *effect* on the behaviour of another. 'Signals' are the means by which these effects are achieved. However, various refinements are then introduced to the definition to distinguish communication from other sorts of effects that animals may have on one another, such as direct physical violence or prey animals inadvertently attracting the presence of predators. It is these refinements that give such a problematical picture of animal communcation. For example, one common refinement is that communication is to be distinguished from direct physical contact by its 'economy of effort' when compared to, say, pushing, shoving, or biting. Cullen (1972) wrote ' . . . to a man the command "Go jump in the lake" is a signal, the push which precipitates him is not'.

In other words, achieving ends by force is not communication, achieving them by making a sound or a gesture is. Dawkins and Krebs (1978) put the same idea in a slightly different way by saying that whereas force uses the individual's own muscle power, communication makes only minimal use of this, relying for its effect on the muscle power of other individuals. 'A male cricket does not physically roll a female along the ground and into his burrow. He sits and sings, and the female comes to him under her own power. From his point of view, this *communication* is energetically more efficient than trying to take her by force.'

But how much more efficient does something have to be than force before we call it communication? In some cases, the economy of effort is obvious. A bird gives an alarm call and the whole flock takes flight. Here a relatively small signal gives rise to the energetically far greater response of all the birds taking off and moving a considerable distance.

In other cases, the economy of effort, if it is achieved at all, it is much less obvious. Red deer stags challenge each other for the possession of a harem during the rutting season by roaring at each other and often a stag is able to get his rival to retreat without a fight, by simply roaring at a faster rate (Clutton-Brock and Albon 1979). Clearly there is some economy of effort here because fighting is a costly, risky business and roaring less so. But roaring is not without its costs either. Stags that have been engaged in a roaring match are temporarily exhausted, even if they have won. They are unlikely to be able to fight for some time afterwards. There is some 'gain' (Dawkins and Krebs 1978) in the system in that the victor's own muscles did not do all the work of getting the rival to go away. But the victor's muscles have had to work very hard (at roaring), even though the rival eventually moved off under his own steam and did not have to be pushed.

'Economy of effort' is introduced into the definition of communication to distinguish it from overt physical violence but it is clear that it does not enable any hard and fast lines to be drawn. Rather, there is a continuum with small inconspicuous and highly economical signals at one end and overt physical violence at the other. In the middle, and often seeming to be not far from physical violence in the risk and effort they involve are large, loud, or conspicuous signals such as the roaring of the deer.

There are very good evolutionary reasons for this. The deer use the roaring precisely because it is physically exhausting and difficult to do. Only stags that can roar at a high rate have the ability to fight well. A stag that retreats in the face of a stag that can out-roar him is almost certainly avoiding the risk of getting into a damaging fight that he is likely to lose anyway. The roaring is a signal by which the stags assess each other's potential fighting ability and it has evolved to be large and conspicuous because a smaller signal could not so accurately indicate fighting ability (Clutton-Brock and Albon 1979) and would be vulnerable to cheating. If a signal is used for assessment of fighting ability – as a substitute for real fighting, that is – the last thing we should expect is 'economy of effort' (Zahavi 1977). On the contrary, we should expect great effort to be put into the signal because it is only in this way that it pays a rival to retreat without fighting.

We should perhaps, then, try to make a distinction between different sorts of signal: signals of conflict, which are used for assessment and which we would expect to be large and uneconomical on the one hand, and on the other, small economical signals where assessment is not involved, which may be nothing more than a 'conspiratorial whisper' (Krebs and Dawkins 1984). This distinction would be easy if it were not for the fact that the conventional definitions of 'signal' and 'communication' make it all but impossible by having yet another refinement made to their definitions.

This refinement is introduced to deal with the fact that not all 'influences' of one animal on another appear to be real 'communication'. For example, if an insect that relies on camouflage for its protection suddenly moves a leg and, by this behaviour influences the behaviour of its predator so that the predator comes and eats it, most people would not want to say that the insect is 'communicating' with its predator.

To distinguish this kind of 'influence' from true communication, the definition is modified once again: only cases in which the influence from one animal to another is achieved by *specially evolved* signals or displays are to be counted as 'communication' (e.g. Krebs and Davies 1981). 'Specially evolved' signals are recognized by being exaggerated, conspicuous, and very obvious (Tinbergen 1952). Cases where an animal behaves and just happens to influence the behaviour of another are not. On this modified definition, a male duck showing a complex, stereotyped, and gaudy courtship display is 'communicating', whereas the unfortunate insect is not.

Now there is justification for saying that duck courtship is different in an important way from the insect betraying itself by a stray movement. Although both influence the behaviour of another animal (female or predator), it is not in the interests of the insect to gain the attention of the predator whereas it is in the interests of the male duck to gain the attention of the female. There has certainly not been any evolutionary change – any 'ritualization', as Huxley (1966) called it – in the insect case to make the leg movement more conspicuous or more likely to attract the attention of the predator. It is therefore, on this 'specially evolved' definition at least, not a 'signal' and 'communication' is not taking place.

Anxious to arrive at a definition of communication that allowed for the inclusion of duck courtship but excluded insects being eaten as a result of their behaviour, most ethologists have restricted the definition of 'communication' to cases where there is evidence of selection having favoured the doers of the behaviour and made them evolve more and more effective ways of influencing other animals. A 'signal' is therefore not just any behaviour by one animal that alters that of another. Such behaviour can be called a signal only if it has been specially evolved to be conspicuous to enhance its signal function. The 'specially evolved signal' refinement of the definition of communication seems, therefore, to be making entirely plausible and reasonable distinctions between what should and what should not be called 'communication'.

It also, however, brings itself into direct conflict with the equally reasonable first refinement of the definition of communication – that of 'economy of effort'. It is the attempt to apply both of these parts of the definition simultaneously that brings about the definitional problems with communication. On the one hand, communication is said to occur when animals influence each other not by physical force but by 'economy of effort'. On the other hand, communication is said to occur only when animals influence each other by means of specially evolved, ritualized signals, which are recognized by being large and exaggerated, and not at all economical. To put this conflict between the two parts of the definition more succinctly, the most 'specially evolved' signals are those that are least economical in terms of the effort needed to produce them, but it is 'economy of effort' which is supposed to characterize signals.

All this means that, very much against any intuitive idea of what the term may mean, wheeling flocks of starlings or shorebirds are not 'communicating' at all because they do not appear to be using conspicuous exaggerated signals.

There must be something wrong. There must be a way of defining 'communication' in such a way that it excludes the unlucky insect and includes flocks of starlings. If there is, it is not commonly used. Most textbooks stick to the 'specially evolved' criterion, thereby confining the whole discussion of animal communication to a subclass of large and exaggerated signals. Very little attention is paid to the more subtle and unritualized ways in which animals influence each others behaviour. They have simply been left out, by definition, from the study of animal communication.

Now we might think we could get out of this definitional impasse by avoiding the word 'signal' altogether and discussing 'information transfer' instead. If animals are influencing each other, even though by very subtle means, it would surely be reasonable to say that they were 'transferring information' even though it might not be appropriate to say that they were using conspicuous 'signals'? Surely dunlin and starlings transfer information even though their flocks may be coordinated by small unritualized movements? It seems, however, that if we take this apparently reasonable step, we find ourselves up against yet another definitional problem: the innocuous, neutral-sounding term 'information transfer' has unfortunately picked up two quite separate definitions. Ethologists use the term 'information transfer' in two different ways and thereby create an even greater amount of confusion than ever.

The two meanings of 'information transfer'

'Information transfer' has both a technical and an everyday meaning. The technical meaning is derived from communications engineers and their need to have a precise measurement of the amount of information being transferred down a wire, say, or from a satellite. It has seemed a very good idea to apply the precise formulae of engineering to animal communication. It holds out the promise of being able to be objective about a subject that is in constant danger of being misinterpreted by our own subjective views of what animals are doing. It certainly achieves objectivity, but not quite in the way people think it does.

'Information transfer' in the technical sense refers to an increase in the ability to predict what is going to happen next as a result of a given event. For example, suppose I know that you were going to one of four cities – York, London, Glasgow, or Edinburgh – but not which one. Then you tell me that you were going to a Scottish city. I still would not know exactly where you were going. What you had told me would, however, have made me better able to predict. My uncertainty about your destination would have been reduced from 1 in 4 cities to 1 in 2 (one of the Scottish ones). In this sense you would have given me *information*. This information is measured for the sake of convenience (engineers' convenience, that is) not in decimal-based numbers, but in binary-based numbers (0 or 1) or bits. The number of bits of information can be worked out by seeing how many yes–no questions would be needed to get the right answer.

So if, as far as I knew, you were initially equally likely to go to any of the four cities, the number of binary or yes–no questions I would have to ask before you said anything would be two (I could ask whether you were going to Scotland or England and then, having established that, ask which of the two appropriate cities there you were going to). After you had told me you were going to Scotland, I could simply ask whether you were going to Glasgow, and whatever you replied, I would know there you were going. My uncertainty about your destination is reduced from two bits (two yes–no questions) before you say anything, to one bit when you tell me you are going to Scotland and to 0 when you say 'Yes' to 'Is it Glasgow?'.

This guessing game has been made deliberately simple to give the idea of what it means to be able to measure 'information transfer' (Shannon and Weaver 1949). It becomes slightly more complicated when the possible outcomes are not equally likely, if, say, I know from past experience that you are more likely to go to London than to any of the others. But it does not matter to this discussion whether you are familiar with these complications or not.

What does matter is the idea that in all cases, it is only possible to measure the amount of information transferred if we can specify what the possible outcomes are before and after the transfer is supposed to have taken place. We must know how much uncertainty has been reduced and the way of getting to this is to work out how many binary (yes–no) questions would be needed before the supposed transfer and how many after. This gives the 'information content' of the message.

When all this is applied to animal behaviour, it is done in a way which is slightly surprising to many people. What is measured is not how much information is transferred to *another animal*, as might be thought, but how much is transferred to a *human observer* looking at the two animals. Suppose animal A gives a signal; animal B then runs away. This happens a sufficient number of times that we are sure this is a genuine case of A influencing B. B seldom runs away unless A has just given the signal and every occurrence of A signalling leads to B running off. We might think we could measure the amount of information A has transferred to B, but that, of course, would be impossible. We have no idea what B thought before or after the signal – no means of guessing how much his uncertainty was reduced because we have no idea of what his uncertainty was, or is.

The only uncertainty we know anything about at all is that of the human observer who has been recording the behaviour of these two animals over a long period of time. He knows that B has a repertoire of some thirty-two behaviour patterns. He also knows that when A has just made the signal, one of these thirty-two, running away, becomes suddenly much more frequent than before. If A has just signalled, the uncertainty about B's behaviour drops suddenly and it becomes very easy to predict what it will do next. We may not know what information has been transferred from A to B, but we can measure that transferred to the observer. Before the signal he knew only that B would perform one of thirty-two possible behaviour patterns. After the signal, he knows it will be only one particular one. On the simplifying assumption that all thirty-two were, before the signal, equally likely, his uncertainty is reduced from five bits ($32 = 2^5$) to 0.

Another way of putting this would be to say that what we are measuring here is the predictability of response by the receiver, given action by the signaller – a precise measure of the degree of influence one is having over the other. We are not making any inferences about what the receiver is learning about the signaller – whether he is being intimidated or what. It is purely a description of action of one animal and response by another and of how predictably one is followed by the other.

The technical meaning of 'information transfer', then, is synonymous with 'influence' of one animal over another. It covers all cases where an observer, knowing that one animal has just done one action is then in a better position to know what a second animal will do next. 'Information transfer' here would cover the starlings in a flock, fish in schools, red deer roaring, *and* the insect being eaten after moving its legs. If we note that insects are more likely to get eaten when they have just moved, then seeing an insect move gives us information – we know that its chance of being eaten has gone up. We are in a better position to predict what is going to happen next. In that sense, the insect has transferred information to us.

The technical sense of 'information transfer' is therefore considerably broader than 'communication'. 'Information transfer' covers all cases where a human being can better predict the behaviour of one animal knowing what another has just done. 'Communication' as we have seen, is usually taken as rather narrower than this. 'Information transfer' runs the whole gamut of influences that one animal may have on another, with no restrictions as to how this influence is achieved (special signal or inadvertent movement) and no implication as to who is benefiting (the influence may be advantageous to the sender and to the receiver, or to neither or to one and not the other). It does not mean that one animal is intentionally 'telling' another one something. It may or may not be, but all we, as observers looking in from the outside, know is that the behaviour is being affected.

We may, of course, be tempted to say that this is a rather odd way to use 'information transfer', one that runs counter to the ordinary, everyday use. This is certainly true and that is precisely why the term 'information transfer', introduced with the intention of clarifying and simplifying the study of animal communication, has in fact had the opposite effect. It seems so restrictive to say that we can only talk about our own uncertainty being reduced when animals do seem to transfer information to other animals and not just to the human beings who happen to be looking on. Wanting to make this point, many ethologists use 'information transfer' not in its technical sense, but in a sense much closer to the ordinary usage of the term.

'Information transfer' in this other sense does refer to animals giving information to each other. The red deer stags we discussed earlier appear to transfer this sort of information about their fighting ability to each other (Clutton-Brock and Albon 1979). As we saw, the stags assess each other for strength and potential fighting ability. But if we were to say, in a colloquial sense, that the stags were 'transferring information' about their fighting ability, we would be using the term in a quite different way from the technical sense we have just discussed. We would be implying (in a colloquial sense) that a stag was transmitting information that if he *were* to fight he would be able to fight hard and well. We would not imply that he was signalling that he was *going* to attack, which would be the technical, 'ability to predict the future' meaning of 'information transfer'.

Prediction of future events, which is the key to information transfer in the technical sense, is not what is meant by this other more colloquial meaning – a difference which has led to no end of misunderstanding (Dawkins and Krebs 1978; van Rhijn 1980; Hinde 1981).

Here we see the confusion caused by having different definitions of 'information transfer'. On one definition, all signals must involve information transfer. In a technical sense, if there is an effect on another animal's behaviour, then information has been transmitted. Without such an effect there could be no information and no communication. But, depending on the circumstances, the animal may or may not be transmitting information, in a colloquial sense, about its future behaviour or about the environment. We could even be in the curious position of claiming both that an animal was transferring information (technical sense) because its behaviour was altering that of other animals and also that it was not transferring information (colloquial sense) because the receiver animal had learnt nothing we could ascertain, either about the environment or about the sender.

Conclusions

Definitions have had an enormous impact on the way people see animal communication. They have affected what we regard as 'communication' and 'signals' and whether we see communication as involving any sort of transfer of information. In particular, an insistence that a definition of communication must include evidence of signals having been specially evolved has led to an emphasis on large, exaggerated signals, for it is only here that we have any evidence of the 'special evolution' demanded by the definition. It has led to a relative neglect of small subtle signals that may be characteristic of much of the social life of animals. Making large exaggerated ritualized signals is a costly, risky business in that it involves an animal in an expenditure of time and energy as well as making it vulnerable to predation. Large signals are therefore expected to occur only when the alteration in the behaviour of the other animal cannot be achieved by less costly means. Assessment tends to lead to the evolution of large signals. But mutually beneficial 'signals' as in a flock of birds which all benefit by staying together may be so small and muted that some definitions of 'signal' exclude them altogether.

Much of the confusion over 'information transfer', too, appears to have been largely brought about by the way it is defined. Arguments which appear to be over something profound about animal communication (such as whether information is being transferred or not) often turn out to be nothing more than whether a protagonist is adopting the technical or the more everyday meaning of the term.

We are at last ready to turn to the genuine, as opposed to the merely definitional, problems about animal communication. Our somewhat lengthy diversion into the ways people use words has, however, been necessary because it was essential to clear our minds about the various meanings of 'signal', 'communication', and 'information transfer' before discussing some of the genuine problems in the study of animal communication. Animal communication has, over the last ten years, been increasingly placed in the mathematical framework of the theory of games under the name of evolutionarily stable strategies. This has brought clarification of a number of difficult issues, but, as we will see in the next chapter, it has also given rise to some misunderstandings of its own.

Further reading

Green and Marler (1979) give a valuable review of animal communication. Wiley (1983) and Krebs and Dawkins (1984) consider the evolution of communication.

9
Evolutionarily stable strategies

Feral cats (*Felix silvestris*). Photograph by Gillian Kerby.

Evolutionarily Stable Strategies (or ESSs) are a way of formalizing the way in which the behaviour of animals in one generation, particularly through their interactions with other animals, affects the frequency of genes in generations yet to come. Now, as we have seen, any study of adaptation attempts to show how structure and behaviour help animals to survive and reproduce. All adaptation is, as we saw, to do with shifting gene frequencies down the generations. So one of the main questions that people raise is what ESSs say about adaptation that could not be said in simpler, non-mathematical terms.

There is also some confusion about what ESSs are and so we start this chapter with a simple introduction to them followed by a brief description of the different kinds of ESS. Then, having seen that ESS models genuinely do add something to our picture of adaptation, we turn to one area where they seem to have added to rather than cleared up confusion, namely, a curious argument about whether animal communication is 'information transfer' or 'manipulation'.

What evolutionarily stable strategies are

There are two elements to an ESS: one is the idea of a 'strategy'; the other is that strategies can be evolutionarily stable.

A strategy is simple to define in theory but much more difficult to pin down in practice. It is just a specification of what an animal does. 'Remove empty eggshells from own nest' might be a strategy for a black-headed gull. 'Always escalate a fight until injured or opponent retreats' might be a strategy (a Hawkish strategy) for another kind of animal. All that is meant by a 'strategy' is a clear specification of what the animal does. In fact, it does not even have to refer to behaviour at all. 'Produce 50 per cent sons and 50 per cent daughters' could be thought of as a reproductive strategy for a parent, although most of our discussion will centre on behavioural examples.

It is important to emphasize that despite the connotations of the word itself, the idea of a behavioural strategy carries no implication that an animal is consciously working out what it is best for it to do, like a general rationally considering his strategy for battle. Because of such implications, the word 'strategy' is perhaps an unfortunate one. 'Program' or 'phenotype' might have been better in this respect as this would emphasize that what is meant is what an animal does, without any need for rationalizing or thinking ahead. An animal need not be conscious of what it is doing or exhibit any evidence of thinking ahead for it to be said to be following a strategy. But, whatever its shortcomings, the term 'strategy' is firmly ensconced in the literature and we would be quite out of step not to use it.

The detail with which a strategy, conscious or unconscious,

has to be specified depends upon what other strategies are thought to be in competition with it. So, if we were concerned solely with two sorts of gull – those that removed empty eggshells and those that left them lying around the nest – we could talk about 'removers' and 'non-removers' as the two strategies. But if we noticed that some birds removed eggshells immediately after their chicks hatched and others waited until the chicks were dry before removing them, we might have to start specifying removers at 0 minutes, removers at 30 minutes . . .60 minutes, and so on.

Here we encounter once again the problem we met over and over again in Chapter 1: the problem of knowing what alternatives natural selection is choosing between or, to use the somewhat grander terminology of ESSs, of 'specifying the strategy set'. Specifying the set of possible strategies that selection chooses between is indeed a major problem for ESS theories, but it is not a problem unique to ESSs. All studies of adaptation involve, as we saw earlier, saying why one kind of animal is better than another. For example, the question 'why do gulls remove eggshells?' apparently had no element of comparison in it. But as we saw in Chapter 1, an implicit comparison is there, even if we do not realize it. The question does not even make sense unless we can say why the removers do better than some alternative kind of gull. An ESS approach, by demanding a clear statement of what strategies are being considered, merely forces us to be explicit about what is implicit anyway. So far, one up to the ESS approach. Being clear about what one is saying must be better than being unclear.

The second element in an ESS is the idea of its evolutionary 'stability'. Having somehow specified the list of possible strategies that natural selection can choose between, the evolutionarily stable strategy is one which, once all members of a population are following it, has higher reproductive success than any alternative. Once the population consists entirely of animals following this one strategy, it will be stable as far as any change in gene frequency is concerned; in other words, no alternative strategy can be spread through the population. The proviso, 'once all members of a population are following it', is important because it emphasizes one of the major differences between an ESS approach and an 'ordinary adaptation' approach to the same problem.

'Ordinary adaptation' simply looks for the superiority of one strategy over another. If removing eggshells results in more surviving offspring than leaving them in place, then the trait of removing eggshells will spread through the population until all gulls do it, which is what seems to have happened. The ESS approach deals with a major complication that tends to arise and may even prevent any one strategy from being evolutionarily stable. It is often the case that one strategy's success at the expense of another is not a fixed quantity but depends on how many other animals in the population are pursuing the same or different strategies.

For example, the Hawkish strategy, 'Always escalate a fight until injured or until opponent retreats', will be extremely successful if all the other animals in the population are pursuing the Doveish strategy of, 'display but retreat at once if opponent escalates'. The Hawks will go around attacking any animal that stands in their way. Because these immediately retreat, the Hawks will be able to take possession of whatever resource it is they are fighting about and the Doves will lose out. Provided there are not too many other Hawks around, the Hawk strategy will be clearly superior to the alternative (Maynard Smith and Price 1973).

But – and this is the big but – its superiority, when rare, will not enable it to spread through the population and completely eliminate the alternative Doveish strategy, at least not if the cost of being injured in a fight is very high. As Hawks, reproductively successful because of winning so many encounters with Doves, become commoner in the population, they will be more and more likely to meet other Hawks. Fights with other Hawks will be a different matter from the easy victories over Doves. They could result in the death or injury of one or both parties. Doves, on the other hand, never get injured and so, while the Hawks are maiming each other, they will be enjoying a considerable reproductive advantage.

Whether 'Hawk' or 'Dove' is the superior strategy depends, then, on how many other Hawks and Doves there are around. Success is dependent on the frequency of the various other sorts of strategists. In a population of Doves, Hawks have the advantage through winning every encounter. In a population of Hawks, Doves have the advantage through being less likely to get injured. For this reason, neither 'Hawk' nor 'Dove' is an 'evolutionarily stable strategy' in the sense that a pure population of either one could easily be 'invaded' by the other. The superiority of each is dependent on their own strategy not becoming too common.

ESS theory is especially well able to cope with this frequency-dependent element of success. Although neither Hawk nor Dove is an evolutionarily stable strategy in itself, ESS theory enables us to work out what will happen. It predicts that the population will stabilize with a mixture of Hawks and Doves. Figure 7 shows how this is worked out. Each encounter is assumed to have a measurable effect on an animal's fitness, that is, its ability to have viable children. If the animals are fighting about food, and only one of them gets it, then the winner would be able to have V more offspring as a result of eating the food. However, being injured reduced fitness and if an animal is injured and forced to retreat, it will have C fewer children. The changes in fitness or 'payoffs' resulting from the three possible sorts of contest are shown in Fig. 7. The probability that each of the three will take place is, of course, dependent on the relative frequencies of Hawks and Doves in the population. Dove–Dove interactions will obviously be rare when there are not many Doves around.

	Hawk	Dove
Hawk	$\frac{1}{2}V-C$	V
Dove	0	$V/2$

Fig. 7 Pay-offs (gains or losses in fitness) resulting from the four possible kinds of encounters between Hawks and Doves. When two doves meet, they share anything to be gained, but incur no loss. When two Hawks meet, they also share gains, but each may incur a loss (through risk of injury) as well. When a Hawk meets a dove, it takes everything to be gained and the Dove gets nothing.

The success of each strategy depends both on its pay-offs and the chances that it will meet an individual adopting its own or a different strategy. The fitness gain or loss to a Hawk as a result of one contest is called $W(H)$.

$$W(H) = W_0 + pE(H,H) + (1 - p)E(H,D)$$

where W_0 is the fitness it had before the contest, p is the frequency of Hawks in the population, $E(H,H)$ is the expected pay-off of the Hawk's encounter with another Hawk, and $E(H,D)$ the expected pay-off with a Dove. Similarly, the fitness gain or loss to a Dove as a result of one contest is called $W(D)$ and

$$W(D) = W_0 + pE(D,H) + (1 - p)E(D,D)$$

The fitness of the Dove strategy is calculable from the payoff to an individual adopting the Dove strategy against a Hawk $(E(D,H))$ multiplied by the number of times it is likely to encounter a Hawk (p) plus the payoff to an individual adopting a Dove strategy against a Dove $(E(D,D))$ multiplied by the likelihood of encountering another Dove $(1-p)$. This is more succinctly put in terms of the equation:

$$W(D) = W_0 + pE(D, H) + (1 - p)E(D, D)$$

where $W(D)$ is the fitness of the Dove strategy. The term W_0 is inserted into the right-hand side of the equation because, before a contest, all individuals are assumed to have a baseline fitness W which changes up or down as the result of their fight. A similar equation can be drawn up for Hawks. As we have previously seen intuitively, Dove is not an ESS because $E(D,D) < E(H,D)$. In other words, a population of Doves will be invaded by Hawks.

But our previous intuition about Hawks – that they too could not form an ESS – was not entirely correct. Hawk is an ESS if what is gained from a fight with another Hawk is greater than the loss through injury. If $V > C$, then a population of Hawks will be stable and uninvadable by Doves. (Note how the ESS approach has cleared our thinking on this point.)

If what is to be gained from victory is less than the risk of injury, that is $V < C$, then the evolutionarily stable mix of strategies can be calculated. This is done by defining the stable point as that mixture of strategies where neither one has an advantage:

$$pE(H, H) + (1 - p)E(H, D) = pE(D<H) + (1 - p)E(D, D)$$

It turns out, remarkably simply, that a population with a frequency of $p = V/C$ Hawks is stable. The stable point is determined by the ratio of the fruits of victory to the cost of defeat. At this stable point, neither Hawk nor Dove is better. Both are exactly equal, given the penalties and advantages each carries and the probability that they will encounter another animal pursuing the same or a different strategy. Such a situation is referred to as a mixed ESS. Whenever the costs of fighting are high relative to what is to be gained, mixed ESSs can be expected, a point we will take up in the next section.

The conclusions of ESS theory are based on certain simplifying assumptions that at first sight are a bit startling. They are based on the assumptions that individuals reproduce asexually, in numbers proportional to their fitnesses, and that the population of which they are a part is infinite in size with random mixing between the strategies (Maynard Smith 1982b). The constraints of real animals and real populations are swept aside. The obvious prevalance of sexual as opposed to asexual reproduction, the niceties of different modes of inheritance are assumed not to matter. A direct link is forged between one generation and the next, giving us a simplified but extremely useful tool. More complex situations, as where individuals tend not to move away from their birthplace and so keep meeting their relatives, do sometimes have to be faced. But often, the simplifying assumptions made by models, such as the one we have just discussed, take us much further towards an understanding of evolution than would have been possible if we had waited until all the facts about pleiotropy and methods of inheritance of aggressive behaviour had been discovered.

There is one other small point to be cleared up. It is often stated that pure ESSs are those which, if all members of a population adopt them, cannot be bettered by any mutant strategy. This is true but slightly misleading. A completely new mutant might well come along and oust an ESS from its position of pre-eminence. But it would have to be outside the strategy set that had already been considered. An ESS is king only within a specified strategy set. It is possible to work out what will happen between Hawks and Doves only so long as these are the only strategies or mutants around. If a completely new strategy is considered, say an animal which appeared 'Doveish' and was never the first to escalate a fight but would stand and fight if provoked, the whole situation is changed. The evolutionary results may be radically altered just as they would be if a new sort of gull, say one which had eggshells that were camouflaged on the insides suddenly arose. The advantage that a 'remover' has over a 'non-remover' when both have conspicuous white eggshells might be completely upset by the arrival of a mutant with camouflaged eggshells and no need to remove them.

There is simply no way in which we can get away from specifying the list of alternatives that we think natural selection chooses between or has chosen between in the past – not if we want to understand the nature of adaptation. The ESS approach demands it, and the discipline of having to think about alternatives is a good one, even though it may show up our ignorance about what those alternatives are or have been. In addition, the ESS approach is a way of dealing with the widespread phenomenon that which strategy is the 'better' may vary with how common that and other strategies are in the population. It is not enough for one strategy to be better than another when it is rare. That will not enable it to spread through the population and become the only variant if it is penalized by its own success. A 'pure' strategy is one that can cope with its own success. Many strategies that are favoured when rare can achieve only a partial spread through the population, sharing the gene-pool with another 'mirror-image' strategy that thrives and falters in a complementary way. The result will then be an evolutionarily stable mixture of strategies.

By the end of this chapter, it will be apparent that such situations are extremely common and that everywhere we look we will find cases of the success of a strategy varying with its frequency relative to some other strategy. The idea of evolutionary stability turns out to be one of the most important pieces of mental equipment we have for dealing with them.

Different kinds of ESS

So far we have discussed two sorts of situation. One is where there is a true 'pure' ESS, a strategy that when all members of a population are following it, cannot be bettered by any of a specified set of alternative strategies. Hamilton (1967) called this an 'unbeatable' strategy, which is a helpful term for the same idea. Then there are mixed strategies – cases where there is no one ESS but where stability in the population is achieved by a mixture of two or more strategies. Any deviation from the stable point is automatically corrected by the fact that the successes of the two strategies are frequency dependent. So, if strategy A becomes commoner, its success goes down over the generations until its frequency is returned to the stable point, and so on.

In the last section, we were deliberately vague about exactly how this stable mix of strategies is achieved. There are, being more precise, two distinct possibilities. Either there could be two sorts of individuals each pursuing one strategy throughout their lifetimes – a genetic polymorphism of Hawks and Doves, for instance. Or there could be one kind of individual that sometimes acted as a Hawk and sometimes as a Dove. Both of these routes could result in the population reaching an evolutionarily stable state, but in the first case there would be a genetic difference between the two kinds of individuals and in the second there would not. In the non-genetic case, an individual would on some occasions be a Hawk and on others the same individual would be a Dove. On a given occasion, whether it would be a Hawk or a Dove would be decided effectively at random.

It is sometimes stated that there is no difference between these two ways of achieving a mixed ESS, that, in other words, a genetic polymorphism with two sorts of individual is exactly equivalent to one sort of individual able to follow two sorts of strategy. We have to be a little careful with such a bland assertion. To begin with, it could only be true if the one sort of individual was as good at pursuing two strategies as genetically different individuals are at pursuing their separate ones. If an individual that was a Hawk for all of its life was better at fighting than an individual that spent some of its time being a Dove and was a Hawk only occasionally, then the equivalence would clearly break down.

From a mathematical point of view, too, the equivalence holds only for situations where there are two strategies and even then not always. With three or more, the mathematics for a genetic polymorphism type of stability and that for a single morph following different strategies breaks down even more often. It becomes important to distinguish between the two because a mixed-individual ESS may be stable in situations in which a mixed population is unstable and vice versa (Maynard Smith 1982b).

But there is a more fundamental problem with mixed ESSs: to conform to a mixed ESS, the behaviour of an animal must contain a random element. Where an individual is sometimes a Hawk and sometimes a Dove, to call this a mixed ESS implies that the decision to do one or the other is random – not in any way adjusted to the circumstances the animal finds itself in. It should not adjust its behaviour to the situation and decide, say, to be a Hawk when it confronted an animal smaller than itself and a Dove if its rival were bigger. If it does make such adjustments, it is not, strictly speaking, following a mixed ESS. It is probably following a 'conditional' strategy, which we will discuss in more detail in a moment. This distinction can be very confusing but it is very important to try and grasp it. The Hawks and Doves we have just discussed had no prudent adjustment to circumstances. With mixed strategies, there is a random element, the equivalent of tossing a coin to decide whether the animal is a Hawk or a Dove in a given encounter. Everybody is kept guessing and the payoffs are calculated on the assumption that all that is known about a rival is that it conforms to the population probability of behaving in one way or another. This does not, on the face of it, appear to be a very sensible way to go through life. An animal that behaved not at random, but by adjusting its behaviour to what it perceived in the environment on a particular occasion would seem to have a clear advantage. It would not get itself into fights with animals that were likely to beat it and it would have a go when faced with a puny animal that it was likely to beat. The 'random element' animal, on the other hand, would find itself fighting all sorts of rivals – both bigger and smaller as well as playing Dove to an animal that it would have no difficulty in beating.

Not surprisingly, then, we find many examples of animals behaving 'sensibly', that is, of adjusting their behaviour to suit the circumstances. These are the strategies that are called conditional strategies. A conditional strategy might be something like 'be a Hawk if rival is bigger but be a Dove if rival is smaller'.

Conditional strategies involve the animal picking up cues which are good indicators of the likely outcome of an encounter, for example, whether the rival is bigger and stronger than it is or likely to fight very hard because it is defending a nest with offspring. These cues are known as 'asymmetries' and they can be of various sorts, known to both contestants or only one and, most importantly, they can be immediately obvious or apparent only after a period of assessment. The roaring of red deer, which we discussed in the last chapter, is a good example of where an asymmetry (in fighting ability) is reflected in a signal that enables the animals to assess each other before a fight.

As we saw then, the trouble with these signals of asymmetry is that they are very vulnerable to cheating. If animals were to use, say size, to decide whether to attack or retreat without fighting, then a weak animal that pretended to be large by puffing itself up would be at an advantage. It would get its rivals to go away even though they might have been capable of beating it. In such a situation, the advantage of using size as a cue to decide whether or not to fight begins to disappear. Animals may then be better off ignoring the size cue if large body size can be mimicked by small animals puffing themselves up.

To evolve as an assessment cue, therefore, an asymmetry has in some way to be cheat-proof, that is to be a genuine indication of fighting abililty. The asymmetry should have the properties of being clearly perceivable by both contestants before escalation occurs and of being costly and expensive to give itself.

Because of the costs of actually fighting and getting killed or injured as a result, we would expect animals to make use of asymmetries if they are available. This raises the question of why animals do not always use conditional strategies, why, in other words, we have even considered the possibility of mixed strategies at all. The Hawk with the random element (mixed ESS) would seem to be at such a disadvantage compared with the Conditional Hawk that it is tempting to think that mixed strategies would always be replaced by conditional ones. In individual combat, they probably are, for the reasons we have discussed, but where it is quite impossible for the animal to judge who or what its rival is going to be, mixed strategies may persist. For example, in considering whether an individual should have male or female offspring, it is impossible to judge exactly what sexes of other adults those offspring will come across when they themselves grow up. The parent here is not competing against an individual opponent with a known percentage of male and female offspring but effectively against the whole population of other parents or at least that section of it whose offspring its own are going to meet. Such a situation is known as 'playing the field' and it is in this sort of case that mixed ESSs seem most likely, with conditional strategies being more characteristic of paired contests between particular individuals where communication and assessment are taking place (Maynard Smith 1982b).

The distinction between mixed and conditional strategies should now be clear. Conditional strategies are pure strategies. All animals in a population are following the same single strategy such as 'be Hawk if rival is smaller, be Dove if rival is bigger', even though we may see the animals behaving sometimes as Hawks and sometimes as Doves. Large animals behave as Hawks much of the time and they may be much more successful at reproducing than the smaller animals that spend most of their time behaving as Doves and have to make 'the best of a bad job'. The behaviour of 'being a Dove when small' can be maintained in the population even though being a Dove is not very successful. The reason for this is that 'be a Dove when smaller' is the junior partner in the highly successful duo 'be a Hawk when larger, be a Dove when smaller'. This duo is the one conditional strategy and its success is judged by that of all its followers taken together.

With mixed ESSs, on the other hand, there are two separate strategies. Hawks and Doves in a mixed ESS are alternative and competing strategies, not part of a single pure strategy held together by an 'if', as they are in a conditional strategy.

The value of ESS models

Critics have said that the ESS may be a mathematically interesting idea but that it is not of very much practical use because of the difficulties of measuring the various costs and benefits involved. It is certainly true that measuring the increase in fitness an animal derives from winning a fight or the decrease it suffers as a result of losing it is a tall order. It is difficult enough simply to show that animals doing one kind of behaviour have some lifetime reproductive advantage over animals doing something else. Showing precisely how many offspring, or parts of offspring, an animal gains or loses from a fight would seem to be all but impossible and yet this is what is demanded by the Vs and Cs in the payoff matrices we have looked at. Without them, it is impossible to calculate the equilibrium frequencies and so, the criticism goes, the point of the mathematics of ESSs is rather lost.

There are two answers to this criticism. The first is that, difficult though it is, it may sometimes be possible to measure the required values. The golden digger wasp, *Sphex ichneumoneus*, lays eggs in underground burrows which are dug by the wasps themselves. The care of each egg consists of provisioning a burrow with katydids as food for when the larva hatches, laying an egg, and then filling up the burrow. A female will complete several such cycles in her brief summer lifetime and by careful observation it is possible to ascertain exactly how many eggs each female has laid. Sometimes a wasp will dig her own burrow but sometimes she will make use of the burrow already dug by another wasp. There thus appear to be two strategies, 'digging' (own burrow) and 'entering' (someone else's). 'Entering' is a successful strategy if the entered burrow is empty because the wasp is saved the time and effort of digging her own burrow. But it is unsuccessful if there is already a female in possession because a fight may ensue and the second wasp may not lay her egg at all. Brockmann, Grafen and Dawkins (1979) found that the average success rate, in eggs laid per 100 hours, was 0.96 for 'digging' and 0.84 for 'entering': It is clear that the success of each strategy is frequency dependent. The more diggers there are in the population, the more burrows there would be around and the higher the success of the entering strategy. But the more frequent enterers there are, the fewer empty burrows there would be and the lower the consequent success of entering. Digging and entering thus seem to form a mixed ESS; the success rate of each strategy could be measured and corresponded to the observed frequencies of the two strategies in the population.

The second and perhaps more important answer to the charge of the impossibility of measuring all the things that ESS theory requires is, that important conclusions can often be reached even though precise numerical values may not be known. We have already seen that in symmetric contests between Hawks and Doves, where the costs of being injured or losing are large compared with the advantages of winning, no pure ESS can result. A mixture of strategies is the only outcome that will be stable.

Another theoretical prediction which can be made in the absence of any precise measurements is about situations in which mixed ESSs would not be expected to arise, indeed simply could not be stable at all. In a contest where both participants perceive an asymmetry, Selton (1980) showed that no mixed strategy can be stable. Assessment and choice of whether to escalate on the basis of information received during the assessment period is favoured even if there is a cost to doing the assessing. In other words, the strategy of 'assessor', which chooses Hawk when larger and Dove when smaller, will be stable against either Hawk or Dove. If there is a cost to escalating the fight, assessor will be the one pure strategy that remains. The only way in which its stability can be breached is if the assessment is not entirely accurate.

ESS models can, then, be a useful way of thinking about a situation even though the increases and decreases in fitness may not be known in detail. Animal communication is one area of ethology that has been particularly helped by ESS models, although it has to be said that their contribution has not been an entirely unmixed blessing. They have clarified in some cases and obscured in others.

'Information' and 'manipulation'

The role of ESS modelling has not been particularly happy in a recent controversy over whether animal communication should be seen as 'information transfer' or as 'manipulation' (Dawkins and Krebs 1978). As we have seen, early ethologists argued that animal signals are often large and exaggerated so that they could better convey information to the animals that were to receive them. Dawkins and Krebs countered this argument by saying that it was often *not* to the advantage of an animal to convey information and that therefore it was impossible to explain the evolution of all ritualized signals in this way. We will try and see our way through this confusing argument one step at a time, starting with an explanation of why Dawkins and Krebs claimed that some animal signals do not convey information at all: on the face of it a quite extraordinary claim to make about animal communication.

The claim came from applying a particular sort of ESS model to animal communication. This model is known as the war of attrition (Bishop and Cannings 1978) and is a form of mixed ESS model, that is, one in which there is no single pure strategy that is stable.

In a war of attrition, we can imagine two Doves displaying at each other. Neither escalates the fight and overt aggression never takes place. The winner is the one that keeps displaying for the longest time. It out-stares or out-displays its rival. Even the staring or the displaying, however, cost something – otherwise both contestants would persist for ever. The situation can be viewed as a mixed ESS in which there are not just two strategies 'display for a short time' and 'display for a long time' but a whole range of strategies defined by the displaying time that the animals are prepared to put up with. Long displaying animals tend to win encounters.

At the beginning of a war of attrition, it is assumed that there is nothing to tell either protagonist how long its rival is prepared to display for, nor what would happen if the encounter did, by any chance, escalate into real fighting. Under such circumstances, it would not pay an animal to convey the information that it was about to retreat, because its rival would then be favoured if it persisted just a little longer. If the rival had itself been on the point of giving up, it could win the encounter by adjusting its behaviour in accordance with the information it had just received. Conveying information about intention to retreat could enable a rival to win and would therefore be selected against.

For this reason, Dawkins and Krebs argued that 'information transfer' cannot possibly be seen as a universal feature of animal communication. Now, there are three quite separate confusions surrounding this argument. The first concerns the different meanings of 'information transfer' we took so long to disentangle in the last chapter. The second concerns whether the war of attrition is the right ESS model to apply to animal communication. The third is an embarrassing obstacle to measuring information transfer between animals.

We have already seen that there is both a technical and a colloquial use of the term 'information transfer'. In the technical sense, all animal communication involves information transfer, by definition. If a human observer sees that one animal's behaviour is having an effect on that of another, then information is necessarily transferred to the observer. If there were no drop in uncertainty – no information transfer – as the observer watched the two animals behave, there would be no sense in which communication could be said to be occurring.

But 'information transfer' in the more colloquial sense may have occurred. An animal may transfer information about its fighting ability (colloquial sense of information) but not about exactly when it is going to attack (technical sense of information). We may be able to predict from its signal how good it would be at fighting if it were to fight, but quite unable to predict whether it was about to attack or not (Maynard Smith 1982b). No wonder there is confusion about whether signals 'convey information' or not!

An even more serious source of confusion, however, comes from the fact that the war of attrition model predicts that animals should not convey any sort of information, either about what they are going to do next or about any asymmetries such as fighting ability that may exist between them. In other words it says that no communication should take place at all and that neither animal should modify its behaviour in the slightest. For the war of attrition, as we noted in passing, is a mixed ESS, not a conditional ESS. We have already seen that mixed ESSs without assessment are *un*likely to evolve in situations of individual combat. In such situations, we expect to find conditional strategies with assessment. So it is very confusing that ESS theorists, having made this point forcibly, then proceed to apply the mixed strategies of the war of attrition to precisely this situation (e.g. Maynard Smith 1982b).

Their surprise that the war of attrition predictions do not seem to hold is therefore a bit puzzling to the rest of us. Animals do seem to pick up some sort of information about each other during a fight (e.g. Riechert 1978, 1979; Hazlett 1982). If they are conveying information about fighting ability and modifying their behaviour according to their perception of an asymmetry between themselves and their rivals, then we are witnessing the playing out of conditional strategies based on assessment, not of mixed strategies. The war of attrition model is inappropriate.

The third difficulty with the idea of information being transferred when animals interact is a purely practical one, but it has seriously obscured the whole issue to the confusion of almost everyone. Remember, once again, that 'information transfer' in the technical sense is defined by the ability of an observer to predict the behaviour of a second animal having seen what a first had just done. 'Information' is transferred to the observer and, by implication, to the second animal.

But suppose the second animal is actually wiser than the human observer. Suppose the second animal sees the first one performing a particular kind of behaviour (a signal) and, having previously learnt that this presages attack, immediately retreats. The signal is such a good predictor of immediate attack that there is no point waiting around. The human observer, however, sees the signal but no attack follows. He therefore concludes that the signal is not a good predictor of attack at all. Discussions of information transfer can be bedevilled by this problem (van Rhijn 1981).

All in all, the question of whether animal signals 'convey information' is one of the most tangled we have come across so far and there is no simple answer to it. 'Information' itself has two different meanings (Krebs and Dawkins 1984), the mixed-ESS models that have been used may be inappropriate when applied to interactions between individual animals, and, as if this were not enough, there are practical difficulties in deciding what, if any, information has been passed between animals. The only thing that is certain is that selection favours animals that leave descendents and sometimes this will involve them in altering the behaviour of other animals to their own advantage, a situation that Krebs and Dawkins called 'manipulation'. Whether they achieve this by conveying or by witholding information will depend partly on the selection pressures involved (specifically, on whether the exchange of information is mutually beneficial to the animals involved or benefits only one of them) but very largely, as we have discovered, on the precise definition of 'information' that is being used.

Conclusions about ESS models

Most of this chapter has been about fighting or contests where physical damage is, or is potentially, inflicted on another animal. The reason for this is probably that these obvious contests, with their roaring or singing or clashing of horns, are the ones that come most readily to our attention. But ESS models can be applied anywhere there are conflicts of interest, that is, wherever the fitnesses of the two parties go up and down in different ways. Siblings in a nest could have a conflict of interest in the sense that if one begs and gets a little more food, its fitness is increased while those of the other nestlings is decreased. But all it does is beg a little harder. It does not fight or threaten its siblings, but there is still an underlying conflict.

Another example of covert conflict are the 'caller' and 'satellite' male crickets (Cade 1981) we met in Chapter 5. 'Caller' males sing for females but also inadvertently attract parasitic flies. 'Satellite' males sing less themselves and try to intercept females on their way to the calling males. Satellites are less effective at getting females but at the same time they are less often attacked by parasitic flies. This situation can be viewed in terms of ESS theory even though callers do not fight satellites. The advantage to each strategy is, moreover, frequency dependent (the more callers, the greater the advantage to satellites and vice versa) and so ideally suited to an ESS analysis.

Anywhere there are conflicts of interest – between parents and young, between chorusing tree frogs, or even between apparently harmonious birds in a flock – the conflict can be analysed as an ESS problem. The ESS approach makes certain simplifying assumptions about the actions of an individual and the frequency of genes in the next generation and so tries to establish a direct connection between the two. For all the difficulties and confusions that sometimes surround them, ESS models are potentially great clarifiers of our view of the adaptiveness of behaviour.

But, of course, fighting, getting enough to eat, and evading predators are only preliminaries. They may help the individual animal to survive but survival of the individual is not in itself what affects gene frequency. Shifting gene frequencies come about through the differential abilities of their carriers to produce viable offspring. We have not so far considered the strategies of mating and reproduction. Yet, to use the analogy from the mathematical theory of games which gave rise to ESS ideas, our understanding of behaviour and evolution would be quite incomplete if we had failed to look at the way animals play out this most serious and important game of all.

Further reading

Dawkins (1980) gives a non-mathematical and Parker (1984) and Maynard Smith (1982b) more mathematical accounts of evolutionarily stable strategies. Maynard Smith's (1984b) article has a variety of views expressed about it in the accompanying commentaries.

10
Sexual selection

Peacock (*Pavo cristatus*). Photograph by Nigella Hillgarth.

Sex is a difficult and much misunderstood subject. There are things about it which nobody fully understands, such as why it is such a widespread method of propagation or why male animals may adopt sexual ornaments so elaborate as to endanger their own survival. There are also some very widespread misconceptions about it, mistakes rather than genuine mysteries.

Both the mysteries and the mistakes will form the basis for this chapter.

The advantages of sex

It has been a fundamental idea in this book that genes are passed from one generation to the next in proportions that are affected by the behaviour of the animals carrying them. We see animals feeding, protecting themselves against predators, or building nests, and we can see all these activities as part of that process. All, that is, until we look at the final generative act. When animals get to the point of making a new generation, a most extraordinary and counterintuitive thing happens.

Instead of the genes that have served them well enough in their own lives all being passed into an offspring, only half of them go. The rest of the offspring's genes are made up by those *from another animal*. The offspring are thus semi-aliens, not flesh of their parent's flesh but only half so, the other half coming from a completely different individual.

From the genes' point of view, the situation is even odder. Engaged, so we had thought, in a bitter struggle with their alleles for a place in the gene-pool, they appear to step aside at the last minute, accepting a 50:50 chance of entering a gamete instead of the certainty they could have had with asexual reproduction in which all the genes are passed on. Because sexual reproduction involves the halving of the adult's chromosome number to make haploid gametes, only one of an allelic pair can enter each gamete. The fact that the two alleles of a diploid adult temporarily cohabit in the same body does not alter the fact that, as far as gene frequencies are concerned, we would expect them to be rivals.

Now, if the two rivals each entered a gamete, even though a different one, and each of these gametes mated and produced an offspring, there would be no real mystery. Each would have much the same chance of becoming part of the reproductive adults of the next generation. There would be a slight problem explaining why they did this in two separate bodies instead of reproducing as a pair, asexually, by duplicating the body they had both been successfully inhabiting for a whole lifetime. But as far as we could tell, both would set off with their new partners with roughly equal chances of success. This is only true, however, if each gamete contributes equally to the nutrition and care of the resulting offspring.

As long as each haploid gamete from one parent mates with another from a second parent, and each gamete is provided with half the nutrients needed to start an embryo, the result is twice as many fertilized eggs as any one parent could manage on its own. The single asexual parent, by contrast, would have to supply its eggs with all their nutrients. The sexual parent would have to supply them with only half and could therefore have twice as many for the same outlay. And a gene with a 50 per cent chance of getting into a particular number of eggs is just as well off as a gene with a 100 per cent chance of getting into half that number. The halving of the chromosome number in gamete formation is not therefore a problem in itself, as is sometimes suggested.

The real problem comes when there are not enough gametes produced for each allele to go its separate way, that is if the number of eggs is not doubled to make up for the drop from 100 per cent to 50 per cent in the chance of entering a gamete caused by the parent making haploid rather than diploid gametes. Yet this is precisely what seems to have happened. Sexual females can make only the same number of eggs as asexual females because the male gametes – the sperm – usually contribute nothing in the way of nutrition at all.

Parker, Baker, and Smith (1972) produced a theory of how this probably occurred. They imagine a primeval situation in which parents produced equal numbers of equally well-provisioned gametes and shed them into the sea. There would have been no small-gamete parents and large-gamete parents as there are now with our 'males' and 'females'. Then they postulated a mutant parent that provisioned its gametes slightly better than any of the others. It might have had to make fewer gametes because of giving each one more, but these fewer well-nourished gametes would have given their subsequent embryos such a head start that the trait would be favoured. Over evolutionary time, some gametes would become bigger and better until some point was reached when extra food would make little difference and the parent would gain more by making another gamete.

While this was happening, other parents, which were still making large numbers of not particularly well-nourished gametes could cash in and effectively parasitize the small number of well-nourished gametes. The embryos could rely on the nourishment by one parent, while the other parent contributed little and put its energy into making lots of other small gametes. Parker et al. argue that this would have led inexorably to the evolution of gametes of very different sizes and our present division into males producing many small sperm and females producing a few large eggs. Female gametes would not mate selectively with other female ones partly because, being fewer of them, they might not find them and partly because they would not gain as much advantage from doing so as a male gamete would. The extra fitness a female gamete would gain from doubling its amount of nutrients would be a lot less than the extra fitness a male one would gain from going from nothing to the female amount. Selection would be heavier on male gametes to seek out and find female ones than on female ones to find other females, coupled with the fact that swimming around to search is probably less easy if encumbered by a large amount of yolk.

The hazard of unequal gamete size is what it does to the numbers of gametes produced. If there were no cost to producing a large gamete, females could produce as many eggs as males produce sperm. Sexual females could produce twice as many eggs as asexual females and make up for the halving of the genes in their eggs by sheer numbers. But there is a cost. The more that is given to one gamete, the less there is for making other gametes. Sexual females simply cannot make up the numbers of their eggs. If they give each one as much yolk as an asexual female gives hers, they make the same number and yet the resulting offspring are only 50 per cent related to them. If the males helped by making a nutritional contribution to each embryo, the females could afford to make larger numbers of less well-nourished eggs, which would still be competitively viable. But the males do not do this. They have evolved to produce more and more smaller gametes that usually make little contribution beyond the bare minimum of genetic material.

And yet sexual reproduction persists! It is a very widespread method of propagation even though genes in female bodies appear not to be gaining twice the number of gametes that they would seem to require to compensate for accepting only a 50 per cent chance of entering a given gamete.

There are some misleading ways in which this so-called 'cost of sex' has been expressed. Note that it is not the halving of the chromosome numbers that causes the anomaly (that could be compensated for by an equal contribution from the male parent enabling the female to double the gamete number). Nor is it unequal gamete size itself, but only the fact that making large gametes exacts a price in terms of the total number that can be produced. The mystery about sexual reproduction is therefore not so much that there is a 'cost to meiosis' *per se*, as is sometimes stated, but a cost that is more correctly and awkwardly described as the 'cost of halving the chromosome number and not being able to double the number of gametes because of the unequal male contribution'. The mystery is what possible advantage there could be to sexual reproduction that makes up for this cost and accounts for its persistence.

This question has caused some embarrassment to evolutionary biologists in the past. Although determined to fit everything into a selective framework and see shifting gene frequencies as the basis of all evolutionary events, they were unable to come up with any satisfactory explanation of the phenomenon that is at once extremely widespread and also appears to suffer from a selective disadvantage of quite enormous (50 per cent) proportions.

Early explanations, that it was 'good for the species' to have the variation that is engendered by the mixing and shuffling of sex, had largely to be discarded. It was an attractive idea to think that sexual reproduction might result in variation staying in populations so that if the environment suddenly changed, the population could 'cope' and have at least some of its members surviving, which a uniform asexual population might fail to do. But by the end of the 1960s, the weakness of such arguments was generally realized. Animals could not evolve or persist if they did things that were detrimental to their own reproduction and beneficial to that of others. Their rivals' genes would under such circumstances spread faster and eliminate the 'altruists' and their tendencies to benefit others. Populations do not harbour deleterious traits against the possibility of future benefit. We cannot expect a trait, to use Sydney Brenner's immortal words, to evolve in the Cambrian 'because it might come in handy in the Cretaceous'.

A trait with a large present disadvantage will quickly disappear. Something other than being 'good for the future of the species' must be invoked to account for its persistence and widespread occurrence.

The embarrassing thing was that nobody could think of anything very convincing. There were some attempts to argue that producing offspring that were different from their parents would benefit the individual parent because the environment might change quickly from one generation to the next, but environments did not seem to change that much, at least not drastically enough to compensate for the apparently massive disadvantage to sex in a constant environment.

Williams (1975) suggested that environments might effectively be thought to be changing if the parent dispersed its young so they ended up in a very different sort of place from the parent. It might be advantageous for the parent to produce many different sorts of offspring, particularly if several offspring landed in one place and competed with one another. 'Sib competition' might favour sex. Bell (1982) took the competition idea even further, but based it not so much on adaptation to an environment changing over time but to an environment which varies over space. He argued that the diverse progeny of a sexual brood will occupy slightly different niches from one another and so compete with one another less fiercely than the uniform progeny of an asexual brood with their exactly similar ecological requirements. More sexual than asexual offspring might therefore result.

There was, generally, during the 1970s, a growing realization of the evolutionary significance of biotic, as opposed to physical factors in the environment and an awareness that they might provide the spatial or temporal change that seemed to be necessary to explain sexual reproduction. Physical environments might remain quite stable but other animals, parasites, and predators could present a constantly changing selective background. Van Valen (1973) saw this as a frantic evolutionary race, each species evolving because the others are doing so, like the Red Queen running hard just to keep in the same place.

Hamilton (1980) has argued that coevolution between parasites and their hosts could provide the necessary 'changing environment' to give a two-fold advantage to animals using sex to reproduce themselves. A host genotype which is very successful at resisting the commonest parasites now will be very much less successful in the future because by that time the parasites will themselves have shifted their own genotype frequencies and the host will no longer be resistant to the commonest parasite type. The only way in which an animal can equip its descendents to deal with the parasites that will probably be around when they grow up is, Hamilton argues, to reproduce sexually with all the possibilities that gives for rapid change.

In 1862, Charles Darwin wrote of the problem of why animals reproduce sexually as opposed to asexually: 'The whole subject is as yet hidden in darkness'. In 1984, John Maynard Smith wrote (1984a): 'Although sexual reproduction is almost universal, its functional significance is still a matter of controversy'. Hamilton may yet be proved right, but until there is more evidence, the subject remains genuinely problematical.

Selection on males and females

Whatever the advantage of sexual reproduction, some things are quite clear. Males and females do exist and their behaviour is often very different. Males, with their large numbers of small gametes, are potentially capable of fertilizing the eggs of many females with their larger, but restricted number, of eggs. Females are thus more limited in their reproductive possibilities than are males. But, of course, gamete size is not the only difference we see. Developing embryos are not just given prepackaged food by their parents. They are often nurtured and protected through a long period of infancy. The time and care invested in offspring after fertilization may be considerable and because they are considerable, they are costly in terms of the total number of offspring that can be made. If a parent nurtures one offspring for a long time, it is effectively prevented from having another for much longer than a parent that gave little to each one and went for quantity rather than quality.

Although there are cases of male parental care and

monogamy in which the parental duties are more or less shared, the female sex tends to be the one investing most in each offspring and taking more 'time off' to rear young than males. In polygamous species, this difference can reach dramatic proportions. The female may do all the nurturing and caring for a relatively small number of offspring while the most successful male may father a very large number of offspring, contributing little to each one except the genetic material contained in the sperm. In elephant seals, almost all of the matings in any one year are performed by just a few males (Le Boeuf 1974) and the same is true of sage grouse (Wiley 1973) and hammerheaded bats (Bradbury 1977). No female in these species could possibly achieve such a high reproductive rate in one season, limited as she would be by the time and effort spent on each offspring.

At this point, two fallacies potentially obscure the account of male and female behaviour. The first fallacy concerns the conclusions that are drawn from the differences in possible reproductive output between the two sexes and the second the kind of evidence that is needed to demonstrate sexual selection.

The exploitation fallacy

It is very easy to assume that males 'exploit' females, 'getting away with' large numbers of offspring at virtually no cost to themselves. From this is but a short step to believing that in some sense it is 'better' to be male than to be female, at least in polygamous species. What this fallacious line of reasoning ignores is that males face many other costs besides the admittedly minimal one of the single sperm that manages to fertilize the egg. To achieve this apparently cheap result, the male will have had to make millions of other sperm which are simply wasted. Then, in order to achieve this mating, the male may have had to fight to gain the female or to defend a territory in a lek to which a female will be attracted. Fighting is often such a costly business that the most successful polygamous males enjoy only a brief reign. The harem masters among elephant seals and red deer have very much shorter reproductive lifespans than the females they, temporarily, possess (Clutton-Brock, Guiness and Albon 1982).

Females may thus invest more in individual offspring than

such males, but the males pay a price in other ways. In fact the greater the reproductive potential of a single male (the lower his parental investment), the higher will be the toll extracted from him in fighting and in competition with other males because they too will be battling to fulfil their reproductive potentials.

An individual male may be reproductively more successful than any female, but for every greater than averagely successful male, there will be other males that are less successful than average. The *average* success of males, as a whole, will be exactly equal to the *average* success of females. If it were not, the sex ratio would shift over evolutionary time towards increasing the frequency of the more successful sex. The observed mixture of males and females in a population was explained by R. A. Fisher (1958) in terms that were typically far ahead of his time. He argued that the sex ratio should be viewed as a mixed ESS (he used the concept but not the terminology) in which a stable equilibrium state would be reached with the advantage to parents of producing offspring of each sex being exactly equal. In a population of females, parents producing males would have an overwhelming advantage and vice versa. The population sex ratio stabilizes when the success of each sex (taken as a whole) is exactly equal.

Males therefore cannot be said to be more successful than females or, as a sex, to exploit them, if 'exploitation' means that the females are at a disadvantage. An individual male may have a more successful reproductive career than any female, but any talk of males exploiting or manipulating the whole female sex smacks of evolutionary instability. Even in a polygamous species, the evolutionarily stable state is the one in which the hard-working females are doing just as well as their more flamboyant harem-holding mates.

Because males in a polygamous species have a much higher reproductive potential than they could ever, all, achieve because of the limited reproductive potential of the females, there may well be competition between the males. Darwin (1871) realized that certain males might have an advantage in this competition if they possessed characters that either helped them to win fights with other males or to attract females to them. Characters that gave males the edge in either of these two ways, Darwin referred to as 'sexually selected'. He specifically excluded from sexual selection any characters that may be useful in reproduction but were not selected by this competition with other males. A male characteristic that helped it to rear young might have an effect on reproductive success, but unless it helped the male to obtain a mate in competition with other males, Darwin did not refer to it as evolving under sexual selection.

The sexual selection fallacy

The second fallacy about the behaviour of males and females that is also quite common is about how the existence of sexual selection should be recognized. It is often supposed that competition between males for females can be inferred solely from the observation that males exhibit a greater *variance* in their reproductive success than females. Krebs and Davies (1981) cite A. J. Bateman's experiment on male and female fruitflies and do just that.

Bateman (1948) put equal numbers of males and females together and counted the number of offspring each produced, using genetic markers to establish parentage. He then showed that the variance of female reproductive success was quite small, that is, almost all the females mated and produced roughly the same number of offspring. He also showed that the males varied much more – some of them had many offspring and others hardly any or none. From this, Krebs and Davies conclude: 'Male reproductive success has a higher variance because of competition between males for mates'. They use high variance in reproductive success as showing, in itself, that there was competition between the males for females and hence sexual selection.

Sutherland (1985) has neatly shown up the fallacy in this line of reasoning. He points out that Bateman's result could arise from random effects with no sexual competition at all. Because the male role in the production of offspring takes only a short time, a male that has just mated can be available to mate again much sooner than a newly mated female with her greater investment of time and energy in each egg. The male has many more opportunities to mate and, as a result of these greater opportunities (for both success and failure), a higher male variance could result from purely random factors.

So, suppose that both sexes searched for mates and mated with whichever member of the opposite sex they happened to meet first. With no fighting between males and no preference by females for one sort of male over another, the result would be higher variance in reproductive success among males than among females. The difference in the time available for searching for mates plus the chance effects of which fly happened to bump into which other one, could have produced Bateman's result.

In a detailed study of damselflies, Banks and Thompson (in press) showed that high male variance in reproductive success did indeed seem to be very largely the result of chance factors in the environment and only weakly, if at all, of sexual selection among males. The number of matings a male obtains seems to depend largely on how long he lives – the longer he is around, the more females he will mate with. Their study indicated that the length of life of a male is random – that is death is a sudden, bolt-from-the blue event that is as likely to strike a younger as an older male. Most females were able to mate but there was little evidence of sexual selection among the males despite considerable variance in male reproductive success.

High variance in reproductive output, or differences in this between the sexes do not, then, provide a guarantee that sexual selection is taking place. It is only if it can be shown that at least some of the male variance is due either to competition between males or to a predilection of females for mating with one sort of male rather than another, that we are entitled to conclude that sexual selection is at work. And that, as one may imagine, is quite difficult to do.

Measuring variance in lifetime reproductive success in a population is bad enough. Sorting out why that variance exists is even worse. Even if that is possible, there remains a puzzling enigma, one that we have met before in earlier chapters. Sexual selection, like other kinds of natural selection, would be expected to use up the variance that it needs to operate. So if certain kinds of males are favoured through competing with other males or through female preference, we would expect that one kind of male to dominate the population. After a while the variance among males would be used up, because only the successful sort of male would be reproducing. The variance on which sexual selection might have operated in the past might be expected to be eliminated by the process of sexual selection. So we might actually find it very difficult to get evidence of genetic sexual superiority among males unless we had at the same time an explanation of why that genetic variance should still persist.

Theories of female choice

Sometimes females choose males on the basis of material possessions such as food or good nesting sites (e.g. Thornhill 1976). But sometimes it appears that females are attracted to some attribute of the males themselves. Darwin (1871) proposed that elaborate male ornaments such as long bright tail feathers in certain bird species evolved to attract females even though they might be positively detrimental to the survival of the males themselves. If the reproductive prize was to be attractive to large numbers of females, then he thought that selection would favour males that had a brief glorious career over ones that were more sensibly adapted to survival but failed to get many matings. Central to Darwin's theory was the idea of female choice: what females were attracted to would determine how the males evolved. This element of sexual selection led many people to be sceptical about the theory itself. Female fancy seemed too whimsical a thing to explain the evolution of traits that could actually threaten male survival.

R. A. Fisher (1958) showed how this might come about. He argued that once a majority of females preferred a particular sort of male, however bizarre his appearance, then other females would be favoured if they chose that sort of male to mate with because they would then have sons that would be attractive to a lot of females. Females choosing in the same way as the majority of other females would 'cash in' on the male's success through having sons like him.

Fisher was proposing frequency-dependent selection with a somewhat different result from the one we discussed in the last chapter. There we discussed cases where the success of a strategy became *less* as the strategy became commoner and stability was reached at the point where success was zero relative to the alternative. In Fisher's theory of sexual selection, there is also a frequency-dependent element, but the success of a strategy becomes greater as it becomes commoner. The more common a particular kind of male becomes, the more advantage accrues both to the males and to the females preferring them. The system 'runs away', success increasing until finally halted by the influence of ordinary natural selection. If a male becomes so bright and colourful that he is eaten by a predator before being able to reach any females, he will obviously be selected against, despite the enthusiasm with which he would be greeted by the females if he were able to survive that long.

More recently, Lande (1982) and Kirkpatrick (1982) have produced genetic models of Fisher's originally verbal argument and have shown that, as usual, Fisher was right. This sort of sexual selection does produce the 'runaway' effect. The genes for producing exaggerated adornments in males and those for giving the preference for such adornments in females will come to be found in the same body, because these are the males and females that will mate. The offspring will therefore inherit the adornment from their fathers and the preference for that adornment from their mothers. Whichever sex the offspring turns out to be, both sets of genes will be favoured, one through being expressed, the other carried along by the succes of the set which is expressed. Each feeds on the other's success and the two together sweep through the population. Fisher's theory shows how, despite the threat to individual survival that male adornments may pose, such characters could evolve. But it is not the only possible explanation.

As an alternative, Zahavi (1975) put forward what he called the 'handicap principle' – the idea that females choose males with elaborate adornments because those males must be extremely physically fit and good at escaping from predators, otherwise they would not have been able to survive. A male with a 'handicap' such as a very long tail is effectively demonstrating that he must have great physical prowess in order to cope.

Criticism was initially heaped upon this theory. People ran computer simulations and declared that the handicap principle would not work (Maynard Smith 1976). It tended to be dismissed as a somewhat eccentric but misguided, explanation of female choice. Then Andersson (1982a) pointed out that there were circumstances in which it might be very important. If possession of a handicap were conditional – that is, only genuinely physically healthy males could sport a handicap – then females might use the possession of a handicap to get themselves a strong healthy mate. If strong healthy males tended to sire strong healthy children, the female could increase her chances of a healthy brood by choosing a male with the biggest handicap. It would be an expensive but cheat-proof test of his health and vigour, exactly similar to the large, cheat-proof signals of fighting ability that we discussed in the last chapter. In both cases the signal itself has to be costly to give, otherwise it could be mimicked by an unhealthy, physically unfit male.

Hamilton and Zuk (1982) suggested a more detailed way in which such conditional handicaps might evolve. They pointed out that many of the puzzlingly elaborate displays of male animals involve the showing of skin or bright feathers. Males that were heavily infested with parasites might simply not be able to mount a particularly bright and gaudy display. Their infections would show in poor-quality plumage, skin, or coat. By choosing a male with bright elaborate adornments, therefore, the female might guarantee herself a strong, parasite-free mate and consequently get her strong, parasite-free children off to a good start in life.

It is important to realize that the Hamilton theory of conditional handicaps is a direct alternative to the Darwin/Fisher view of things. On Fisher's theory, female choice is based on whim – the elaborate male adornments are attractive in themselves. They signal nothing else about the male other than that he possesses them. As long as most other females have the same whim, the adornment will persist and evolve. Females will be favoured if they choose adorned males even if those males are otherwise weak and physically unhealthy. They will still gain advantage through having attractive, although possibly not very healthy, sons.

On the Hamilton theory, however, female choice has more substance. The males are being assessed for health and disease-free vigour. The assessment cue which the females are using is no mere arbitrary whim. It is costly for the male and can be sustained only by genuinely healthy males in good condition. By choosing such a male, a female gains strong healthy children of both sexes. On Fisher's theory, by contrast, she does not gain stronger or healthier children – merely male children that grow up to be attractive to females.

We can see that testing these two theories is extraordinarily difficult. It is difficult enough showing that females really are attracted to elaborate male adornments themselves, as opposed to something that the male possesses, such as a territory. Often the most elaborately adorned male will also have the best territory or will have fought off other males so that it may not be at all clear what has been the crucial deciding factor in determining a particular mating.

Andesrson (1982b) performed one of the few experiments showing that male ornament in itself is attractive to females. He cut the trailing tails off male widowbirds and stuck them on again. Some males had longer tails stuck to them, others shorter tails and others, as a control, tails of the same length. Females were attracted to the long-tailed males, even those with another individual's tail stuck to it.

Although this suggests that adornment is attractive to females, it does not say why the preference evolved. Perhaps, in the normal way, long bright tails signify healthy, parasite-free males in which case Hamilton may be right. But perhaps long-tailed males are no more likely to be parasite-free than short-tailed ones and the advantage to the females accrues purely through having attractive male offspring, as Fisher suggested. And, as Partridge and Halliday (1984) point out, competition between males, rather than choice by females, may also be at work.

As with sex itself, we are still almost as in the dark about sexual selection as Darwin was. There are many things that we simply do not understand. Hamilton has suggested that the answer to both lies in parasites – parasites causing a constantly changing environment that only sexual reproduction can keep up with and parasites providing the stimulus for costly, cheat-proof signals before a female chooses a male. Is it possible that two of the greatest puzzles in behavioural evolution will be explained by . . . parasites?

Further reading

Williams (1975) and Maynard Smith (1984a) discuss the various theories for the evolution of sex. Partridge and Halliday (1984) and Bateson (1983b) give reviews of theories of sexual selection.

Epilogue

There may be some people who feel that seeing animal behaviour as a means for passing genes on from one generation to the next somehow belittles or diminishes it. It perhaps makes it seem as though all the things we have discussed – caring for relatives, courtship, fighting – are 'nothing but' the working out of this process. I would like to end this book on a personal note by saying that, for me, nothing could be further from the truth. To give an explanation of something may take away its mystery but it does not take away our capacity to marvel at it. In fact, it may become more marvellous when we realize what has given rise to it.

It does not diminish Abraham Lincoln's achievements to say that he was 'nothing but' a backwoodsman. We admire him and what he did the more when we realize that he did not start from a position of power but that what he did came from his own efforts and personality. His humble beginnings actually make his final achievements more remarkable. So, in an infinitely greater way, the simple beginnings of animal behaviour – the battle of the shifting gene frequencies – make the complexity of what we see animals doing into something truly marvellous.

The fight for a place in the gene-pool has given rise to some of the most beautiful and intricate phenomena on this earth. Animals are not formless masses of jelly, mindlessly reproducing themselves. They have developed the power to swim and to fly, to care for their young, to stalk their prey, to play, to sing and to be curious about the world around them. To know that all this comes from such simple beginnings can enhance and deepen the wonder, not diminish it.

The computer that beats you at chess may be 'nothing but' a load of switches, but what an achievement for the switches – far more remarkable than if there were a man inside the box countering your every move.

The animal that builds a nest and cares for its young may be 'nothing but' the result of natural selection but what a result! From the action of selection, favouring one kind of animal over another down millions of years, have come co-operation, life-long bonds between individuals, and, ultimately, the ability to reason and understand it all. I can think of few things as marvellous or more worthy of a lifetime's study than that.

Fox (*Vulpes vulpes*) investigating radio tracking apparatus used to study foxes. Photograph by Malcolm Newdick.

References

Andersson, M. (1982a) Sexual selection, natural selection and quality advertisement. *Biological Journal of the Linnean Society* **17**,375–93.

Andersson, M. (1982b) Female choice selects for extreme tail length in a widowbird. *Nature Lond.* **299**,818–20.

Banks, M. J. and Thompson, D. J. (in press) Lifetime mating success in the damselfly *Coenagrion puella. Animal Behaviour*.

Barlow, G. W. (1968) Ethological units of behavior. In *The Central Nervous System and Fish Behavior* (ed. D. Ingle). University of Chicago Press, Chicago, pp.217–32.

Bateman, A. J. (1948) Intra-sexual selection in *Drosophila. Heredity* **2**,349–68.

Bateson, P. P. G. (1983a) Genes, environment and the development of behaviour. In *Animal Behaviour*, vol. 3, *Genes, Development and Learning* (eds. T. R. Halliday and P. J. B. Slater). Blackwell Scientific Publications, Oxford, pp.51–81.

Bateson, P. P. G. (1983b) *Mate Choice.* Cambridge University Press, Cambridge.

Bell, G. (1982) *The Masterpiece of Nature.* Croom Helm, London.

Bentley, D. and Hoy, R. (1972) Genetic control of the neuronal network generating cricket (*Teleogryllus*) song patterns. *Animal Behaviour* **20**,478–92.

Bentley, D. and Konishi, M. (1978) Neural control of behaviour. *Annual Review of Neuroscience* **1**,35–60.

Bishop, D. T. and Cannings, C. (1978) A generalised war of attrition. *Journal of Theoretical Biology* **70**,85–124.

Bolles, R. C. (1975) *Theory of Motivation*, 2nd edn. Harper and Row, New York.

Bradbury, J. W. (1977) Lek mating behaviour in the hammer-headed bat. *Zeitschrift für Tierpsychologie* **45**,225–55.

Brockmann, H. J. Grafen, A. and Dawkins, R. (1979) Evolutionarily stable nesting strategy in a digger wasp. *Journal of Theoretical Biology* **77**,473–96.

Brown, J. L., Brown, E. R., Brown, S. D. and Dow, D. D. (1982) Helpers: effects of experimental removal on reproductive success. *Science* **215**,421–2.

Burrows, M. and Rowell, C. H. F. (1973) Connections between descending interneurons and metathoracic motoneurons in the locust. *Journal of Comparative Physiology* **85**,221–34.

Cade, W. (1981) Alternative mating strategies: genetic differences in crickets. *Science* **212**,563–4.

Clutton-Brock, T. H. and S. D. Albon (1979) The roaring of red deer and the evolution of honest advertisement. *Behaviour* **69**,145–70.

Clutton-Brock, T. H. and Harvey, P. H. (1984) Comparative approaches to investigating adaptation. In *Behavioural Ecology*, 2nd edn. (eds. J. R. Krebs and N. B. Davies). Blackwell Scientific Publications, Oxford, pp.1–29.

Clutton-Brock, T. H., Guiness, F. E. and Albon, S. D. (1982) *Red Deer. Behaviour and Ecology of Two Sexes*. Edinburgh University Press, Edinburgh.

Cohen, S. and McFarland, D. J. (1979) Time-sharing as a mechanism for the control of behaviour sequences during the courtship of the three-spined stickleback (*Gasterosteus aculeatus*). *Animal Behaviour* **27**,270–83.

Cowie, R. (1977) Optimal foraging in great tits, *Parus major*. *Nature, Lond.* **268**,137–9.

Cullen, E. (1957) Adaptations in the kittiwake to cliff-nesting. *Ibis* **99**,275–302.

Cullen, J. M. (1960) Some adaptations in the nesting behaviour of terns. *Proceedings of the XIIth International Ornithological Congress*, Helsinki, pp.153–7.

Cullen, J. M. (1972) Some principles of animal communication. In *Non-verbal Communication* (ed. R. A. Hinde). Cambridge University Press, Cambridge, pp.101–22.

Davis, J. M. (1980) The coordinated aerobatics of Dunlin flocks. *Animal Behaviour* **28**,668–73.

Darwin, C. (1871) *The Descent of Man and Selection in Relation to Sex*. Murray, London.

Dawkins, R. (1980) Good strategy or evolutionarily stable strategy? In *Sociobiology: Beyond Nature/Nurture?* (eds. G. W. Barlow and J. Silverberg). Westview Press, Boulder, Colorado, pp.331–67.

Dawkins, R. (1982) *The Extended Phenotype*. W. H. Freeman, Oxford.

Dawkins, R. and Dawkins, M. (1973) Decisions and the uncertainty of behaviour. *Behaviour*, **45**,83–103.

Dawkins, R. and Krebs, J. R. (1978) Animal signals: information or manipulation? In *Behavioural Ecology*, lst edn. (eds. J. R. Krebs and N. B. Davies). Blackwell Scientific Publications, Oxford, pp.282–309.

Dewsbury, D. A. (1978) *Comparative Animal Behavior*. McGraw-Hill, New York.

Dorsett, D. A., Willows, A. O. D. and Hoyle, G. (1973) The neuronal basis of behavior in *Tritonia*. IV. The central origin of a fixed action pattern demonstrated in the isolated brain. *Journal of Neurobiology* **4**,287–300.

Drent, R. H. and Daan, S. (1980) The prudent parent: energetic adjustments in avian breeding. *Ardea* **68**,225–52.

Ehrman, L. and Parsons, P. A. (1976) *The Genetics of Behavior*. Sinauer, Sunderland, Mass.

Ewert, J. P. (1980) *Neuroethology*. Springer-Verlag, Berlin.

Feekes, C. (1972) 'Irrelevant' ground pecking in agonistic situations in Burmese Red Junglefowl (*Gallus gallus spadiceus*). *Behaviour* **43**,186–326.

Feng, A. S., Simmons, J. A. and Kick, S. A. (1978) Echo detection and target-ranging neurons in the auditory system of the bat *Eptesicus fuscus*. *Science* **202**,645–8.

Fisher, R. A. (1958) *The Genetical Theory of Natural Selection*, 2nd edn. Dover, New York.

Frisch, K. von (1967) *The Dance Language and Orientation of Bees*. Belknap Press, Cambridge, Mass.

Gould, S. J. (1978) *Ever Since Darwin*. Burnett, London.

Gould, S. J. and Lewontin, R. C. (1979) The spandrels of San Marco and the Panglossian paradigm: a critique of the adaptationist programme. *Proceedings of the Royal Society of London,* B**205**,581–98.

Grafen, A. (1982) How not to measure inclusive fitness. *Nature, Lond.* **298**,425–6.

Grafen, A. (1984) Natural selection, kin selection and group selection. In *Behavioural Ecology*, 2nd edn. (eds. J. R. Krebs and N. B. Davies). Blackwell Scientific Publications, Oxford, pp.62–84.

Green, S. and Marler, P. (1979) The analysis of animal communication. In *Handbook of Behavioral Neurobiology*, vol. 3 (eds. P. Marler and J. G. Vandenburgh). Plenum, New York, pp.73–158.

Greenburg, L. (1979) Genetic component of bee odor in kin recognition. *Science* **206**,1095–7.

Griffin, D. (1958) *Listening in the Dark*. Yale University Press, New Haven.

Hailman, J. P. (1967) The ontogeny of an instinct. *Behaviour Supplement* **15**,1–196.

Hailman, J. P. (1969) How an instinct is learned. *Scientific American* **221**(6),98–108.

Hamilton, W. D. (1964) The genetical evolution of social behaviour. *Journal of Theoretical Biology* **7**,1–52.

Hamilton, W. D. (1967) Extraordinary sex ratios. *Science* **156**,477–88.

Hamilton, W. D. (1980) Sex versus non-sex versus parasite. *Oikos* **35**,282–90.

Hamilton, W. D. and Zuk, M. (1982) Heritable true fitness in birds: a role for parasites? *Science* **218**,384–6.

Hazlett, B. A. (1982) Resource value and communication strategy in the Hermit crab, *Pagurus bernhardus* (L). *Animal Behaviour* **30**,135–9.

Henderson, N. D. (1970) Genetic influences on the behaviour of mice can be obscured by laboratory rearing. *Journal of Comparative and Physiological Psychology* **72**,505–11.

Hinde, R. A. (1959) Unitary drives. *Animal Behaviour* **7**,130–41.

Hinde, R. A. (1960) Energy models of motivation. *Symposium of the Society for Experimental Biology* **14**,199–213.

Hinde, R. A. (1970) *Animal Behaviour* McGraw-Hill, New York.

Hinde, R. A. (1981) Animal signals: ethological and game theory approaches are not incompatible. *Animal Behaviour* **29**,535–42.

Hoogland, J. L. and Sherman, P. W. (1976) Advantages and disadvantages of Bank swallow coloniality. *Ecological Monographs* **46**,33–58.

Hoyle, G. (1983) On the way to neuroethology: the identified neuron approach. In *Neuroethology and Behavioural Physiology* (eds. F. Huber and H. Markl). Springer-Verlag Berlin, pp.9–25.

Hsia, D. Y.-Y. (1970) Phenylketonuria and its variants. In *Progress in Medical Genetics*, vol. 7 (eds. A. G. Steinberg and A. G. Bearn). Grune and Stratton, New York, pp.29–68.

Huber, F. and Markl, H. (1983) *Neuroethology and Behavioural Physiology*. Springer-Verlag, Berlin.

Huxley, J. (1966) Ritualization of behaviour in animals and men. *Philosophical Transactions of the Royal Society of London* B**251**,249–71.

Kater, S. B. and Rowell, C. H. F. (1973) Integration of sensory and centrally programmed components in generation of cyclical feeding activity of *Helisoma trivolvis*. *Journal of Neurophysiology* **36**,142–55.

Kettlewell, H. B. D. (1955) Selection experiments on industrial melanism in the Lepidoptera. *Heredity* **9**,323–35.

Kirkpatrick, M. (1982) Sexual selection and the evolution of female choice. *Evolution* **36**,1–12.

Kovac, M. P. and Davis, W. J. (1977) Behavioral choice: neural mechanisms in *Pleurobranchaea*. *Science* **198**,632–4.

Kovac, M. P. and Davis, W. J. (1980) Reciprocal inhibition between feeding and withdrawal behaviors in *Pleurobranchaea*. *Journal of Comparative Physiology* **139**,77–86.

Kortlandt, A. (1949) Eine Übersicht der angeborenen Verhaltensweisen des Mitteleuropäischen Kormorans (*Phalacocorax carbo sinensus*). *Archives néerlandaises de Zoologie* **14**,401–42.

Krebs, J. R. and Dawkins, R. (1984) Animal signals: mind-reading and manipulation. In *Behavioural Ecology*, 2nd edn. (eds. J. R. Krebs and N. B. Davies). Blackwell Scientific Publications, Oxford, pp.380–402.

Krebs, J. R. and Davies, N. B. (1981) *An Introduction to Behavioural Ecology*. Blackwell Scientific Publications, Oxford.

Krebs, J. R. and McCleery, R. (1984) Optimization in behavioural ecology. In *Behavioural Ecology*, 2nd edn. (eds. J. R.

Krebs and N. B. Davies). Blackwell Scientific Publications, pp.91–121.

Kruijt, J. P. (1964) Ontogeny of social behaviour in Burmese Red Junglefowl (*Gallus gallus spadiceus* Bonnaterre). *Behaviour Supplement* **12.**

Kung, C., Chang, S.-Y., Satow, Y., Houten, J. and Hansma, H. (1975) Genetic dissection of behaviour in *Paramecium*. *Science* **188**,898–904.

Lagerspetz, K. M. J. (1969) Aggression and aggressiveness in laboratory mice. In *Aggressive Behaviour* (eds. S. Garattini and E. B. Sigg). Excerpta Medica, Amsterdam, pp.77–85.

Lande, R. (1982) Rapid origin of sexual isolation and character divergence within a cline. *Evolution* **36**,213–23.

Le Boeuf, B. J. (1974) Male–male competition and reproductive success in elephant seals. *American Zoologist* **14**,163–76.

Lehrman, D. S. (1953) A critique of Konrad Lorenz's theory of instinctive behavior. *Quarterly Review of Biology* **28**,337–63.

Leyhausen, P. (1965) Über die Funktion der relativen Stimmungshierarchie. *Zeitschrift für Tierpsychologie* **22**,412–94.

Lorenz, K. (1932) Betrachtungen über das Erkennen der arteigenen Triebhandlungen der Vogel [A consideration of methods of identification of species-specific instinctive behaviour patterns in birds]. *Journal für Ornithologie* **80**(1). Translated in K. Lorenz (1970) *Studies in Animal and Human Behaviour*, vol. 1 (transl. R. Martin). Methuen, London, pp.57–100.

Lorenz, K. (1937) Über die Bildung des Instinktbegriffes [The establishment of the instinct concept]. *Die Naturwissenschaften* **25**(19),289–300,307–18,324–31. Translated in K. Lorenz (1970) *Studies in Animal and Human Behaviour*, vol. 1 (transl. R. Martin). Methuen, London, pp.259–315.

Lorenz, K. (1950) The comparative method in studying innate behaviour patterns. *Symposium of the Society for Experimental Biology* **4**,221–68.

Lorenz, K. (1954) Psychologie und Stammesgeschichte [Psychology and phylogeny]. In G. Heberer *Psychologie und Stammesgeschichte*, 2nd edn. Jena. Translated in K. Lorenz

(1970) *Studies in Animal and Human Behaviour*, vol. 2 (transl. R. Martin). Methuen, London, pp.146–245.

Lorenz, K. (1965) *Evolution and Modification of Behavior*. University of Chicago Press, Chicago.

McFarland, D. J. and Houston, A. I. (1981) *Quantitative Ethology. The State Space Approach*. Pitman, London.

McNeill Alexander, R. (1982) *Optima for Animals*. Arnold, London.

Maynard Smith, J. (1976) Sexual selection and the Handicap principle. *Journal of Theoretical Biology* **57**,239–42.

Maynard Smith, J. (1978) Optimization theory in evolution. *Annual review of Ecology and Systematics* **9**,31–56.

Maynard Smith, J. (1982a) The evolution of social behaviour – a classification of models. In *Current Problems in Sociobiology* (ed. King's College Sociobiology Group). Cambridge University Press, Cambridge. pp.29–44.

Maynard Smith, J. (1982b) *Evolution and the Theory of Games*. Cambridge University Press, Cambridge.

Maynard Smith, J. (1984a) The ecology of sex. In *Behavioural Ecology*, 2nd edn. (eds. J. R. Krebs and N. B. Davies). Blackwell Scientific Publications, Oxford, pp.201–21.

Maynard Smith, J. (1984b) Game theory and the evolution of behaviour. *Behavioral and Brain Sciences* **7**,37–44.

Maynard Smith, J. and Price, G. R. (1973) The logic of animal conflict. *Nature, Lond.* **246**,15–18.

Maynard Smith, J. and Riechert, S. E. (1984) A conflicting tendency model of spider agonistic behaviour in hybrid-pure population line comparisons. *Animal Behaviour*, **32**, 564–78.

Miller, N. E. (1957) Experiments on motivation. Studies combining psychological, physiological and pharmacological techniques. *Science* **126**,1271–8.

Morgan, J. T. (1911) Random segregation versus coupling in Mendelian inheritance. *Science* **33**,384.

Morgan, T. H. (1919) *The Physical Basis of Heredity*. Lippincott, Philadelphia.

Mulligan, J. A. (1966) Singing behavior and its development in the song sparrow, *Melospiza melodia*. *University of California Publications in Zoology* **81**,1–76.

Nelson, J. B. (1967) Colonial and cliff-nesting in the gannet. *Ardea* **55**,60–90.

Oatley, K. (1978) *Perceptions and Representations*. Methuen, London.

O'Shea, M. and Williams, J. L. D. (1974) The anatomy and output connection of a locust visual interneurone; the lobular giant movement detector (LGMD) neurone. *Journal of Comparative Physiology* **91**, 257–66.

Oster, G. F. and Wilson, E. O. (1978) *Caste and Ecology in the Social Insects*. Princeton University Press, Princeton.

Parker, G. A. (1984) Evolutionarily stable strategies. In *Behavioural Ecology*, 2nd edn. (eds. J. R. Krebs and N. B. Davies). Blackwell Scientific Publications, Oxford, pp.30–61.

Parker, G. A., Baker, R. R. and Smith, V. G. F. (1972) The origin and evolution of gamete dimorphism and the male–female phenomenon. *Journal of Theoretical Biology* **36**,529–53.

Partridge, B. L. and Pitcher, T. J. (1979) Evidence against a hydrodynamic function for fish schools. *Nature, Lond.* **279**,418–19.

Partridge, L. (1983) Genetics and Behaviour. In *Animal Behaviour*, vol. 3 *Genes, Development and Learning* (eds. T. R. Halliday and P. J. B. Slater). Blackwell Scientific Publications, Oxford, pp.11–51.

Partridge, L. and Halliday, T. R. (1984) Mating patterns and mate choice. In *Behavioural Ecology*, 2nd edn. (eds. J. R. Krebs and N. B. Davies). Blackwell Scientific Publications, Oxford, pp.222–50.

Patterson, I. J. (1965) Timing and spacing of broods in the black-headed gull *Larus ridibundus*. *Ibis* **107**,433–59.

Potts, W. K. (1984) The chorus-line hypothesis of manoeuvre coordination in avian flocks. *Nature, Lond.* **309**,344–5.

Putten, van G. and Dammers, J. (1976) A comparative study of the well-being of piglets reared conventionally and in cages. *Applied Animal Ethology* **2**,339–56.

Rhijn, J. G. van (1980) Communication by agonistic displays: a discussion. *Behaviour* **74**,284–93.

Ridley, M. (1983) *The Explanation of Organic Diversity*. Oxford University Press, Oxford.

Riechert, S. E. (1978) Games spiders play: behavioural variability in territorial disputes. *Behavioral Ecology and Sociobiology* **3**,135–62.

Riechert, S. E. (1979) Games spiders play. II. Resource

assessment strategies. *Behavioral Ecology and Sociobiology* **6**,121–8.

Rose, S. (1978) Pre-Copernican sociobiology? *New Scientist* **80**,45–6.

Rose, S., Kamin, L. J. and Lewontin, R. C. (1984) *Not in Our Genes*. Penguin, Harmondsworth.

Rubenstein, D. I. (1980) On the evolution of alternative mating strategies. In *Limits to Action. The Allocation of Individual Behavior* (ed. J. Staddon). Academic Press, New York, pp.65–100.

Sales, G. and Pye, D. (1974) *Ultrasonic Communication by Animals*. Chapman and Hall, London.

Schnitzler, H. U. (1973) Control of Doppler-shift compensation in the Greater Horseshoe Bat, *Rhinolophus ferrumequinum*. *Journal of Comparative Physiology* **82**,79–92.

Selten, R. (1980) A note on evolutionarily stable strategies in asymmetric animal conflicts. *Journal of Theoretical Biology* **84**,93–101.

Shannon, C. E. and Weaver, W. (1949) *The Mathematical Theory of Communication*. University of Illinois Press, Urbana.

Sibly, R. M. (1983) Optimal group size is unstable. *Animal Behaviour* **31**,947.

Simmons, J. A., Fenton, M. B. and O'Farrell, M. J. (1979) Echolocation and the pursuit of prey by bats. *Science* **203**,16–21.

Simmons, J. A. and Vernon, J. A. (1971) Echolocation: discrimination of targets by the bat *Eptesicus fuscus*. *Journal of Experimental Zoology* **176**,315–28.

Southwick, C. H. (1968) Effects of maternal environment on aggressive behavior of inbred mice. *Communications in Behavioral Biology* **1**,49–59.

Stamps, J. A. and Barlow, G. W. (1973) Variation and stereotypy in the displays of *Anolis aeneus* (Sauria: Inguanidae). *Behaviour* **47**,67–94.

Strickberger, M. (1968) *Genetics*. MacMillan, New York.

Sturtevant, A. H. (1913) The linear arrangement of six sex-linked factors in *Drosophila*, as shown by their mode of association. *Journal of Experimental Zoology* **14**,43–59.

Sutherland, W. J. (1985) Chance can produce a sex difference in variance in reproductive success and explain Bateman's data. *Animal Behaviour* (In press).

Thorpe, W. H. (1961) *Bird Song*. Cambridge University Press, Cambridge.

Thornhill, R. (1976) Sexual selection and nuptual feeding behavior in *Bittacus apicalis* (Insecta:Mecoptera). *American Naturalist* **110**,529–48.

Tinbergen, N. (1951) *The Study of Instinct*. Oxford University Press.

Tinbergen, N. (1952) 'Derived' activities, their causation, biological significance, origin and emancipation during evolution. *Quarterly Review of Biology* **27**,1–32.

Tinbergen, N. (1963) On the aims and methods of ethology. *Zeitschrift für Tierpsychologie* **20**,410–33.

Tinbergen, N., Broekhuysen, G. J., Feekes, F., Houghton, J. C. W., Kruuk, H. and Szulc, E. (1967) Egg shell removal by the Black-Headed Gull, *Larus ridibundus L.*: a behaviour component of camouflage. *Behaviour* **19**,74–117.

Tullock, G. (1971) The coal tit as a careful shopper. *American Naturalist* **105**,77–9.

Van Valen, L. (1973) A new evolutionary law. *Evolutionary Theory* **1**,1–30.

Vestergaard, K. (1980) The regulation of dustbathing and other behaviour patterns in the laying hen: a Lorenzian approach. In *The Laying Hen and its Environment* (ed. by R. Moss). Martinus Nijhoff, The Hague, pp.101–20.

Weihs, D. (1975) Some hydrodynamical aspects of fish schooling. In *Swimming and Flying in Nature* (eds. T. Y.–T. Wu, C. J. Brockaw and C. Brennen). Plenum Press, New York. pp.703–18.

Wiley, R. H. (1983) The evolution of communication: information and manipulation. In *Animal Behaviour* vol. 3 (eds. T. R. Halliday and P. J. B. Slater). Blackwell Scientific Publications, Oxford. pp.156–89.

Williams, G. C. (1975) *Sex and Evolution*. Princeton University Press, Princeton.

Willows, A. O. D., Dorsett, D. A. and Hoyle, G. (1973) The neuronal basis of behavior in *Tritonia*. III. Neuronal mechanism of a fixed action pattern. *Journal of Neurobiology* **4**,255–85.

Wilz, K. (1970) Causal and functional analysis of dorsal pricking and nest activity in the courtship of the 3-spined stickleback, *Gasterosteus aculeatus. Animal Behaviour* **18**,115–24.

Zahavi, A. (1975) Mate selection – a selection for a handicap. *Journal of Theoretical Biology* **53**,205–14.

Zahavi, A. (1977) Reliability in communication systems and the evolution of altruism. In *Evolutionary Ecology* (eds. B. Stonehouse and C. M. Perrins). MacMillan, London, pp.253–9.

Index

Assessment, 95, 103–4, 110–11, 121–3, 144–5
Bats, 14–15, 86–90
Black boxes, 90–8
Cistern model, 70–9
Coefficient of relatedness, 35
Conditional strategy, 121–3
Cost function, 31
'Cost of meiosis', 135
Cost of sex, 132–4
Crickets, 49, 59, 62–3, 129
Developmentally fixed behaviour, 58–62
Digger wasps, 124
Displacement activities, 78–81
Drive, 68–72, 84
Echolocation, 14, 86–90
Economics and animal behaviour, 30
Egg-shell removal, 9, 11, 12, 47
Energy and motivation, 68–73
Escape swimming response, 74–5
Female choice, 142
Fish schooling, 16
Fixed action patterns, 67, 72–6
Genetic determinism, 46, 51
Goal function, 31
Handicap principle, 143–4
Hawks and Doves, 115

Information transfer, 101, 106–13, 125–8
Instinct, 57, 64, 67–72
Jumping in locusts, 85
Kittiwakes, 11
Learning ability of mice, 52
Manipulation, 113, 125, 129
Mixed strategies, 120–3
Motivation, 84, 93–7
Optimal foraging theory, 20, 24–8
Paramecium, 48
Parasites and sex, 136–7, 145
Peppered moth, 7, 8
Phenylketonuria, 52
Psychohydraulic model, 71, 79
Relative
  helping, 35, 47
  recognition, 50–1
Ritualisation, 104, 111
Runaway theory, 142–3
Schooling in fish, 16
Sex, cost of, 132–4
Signals, 101–5, 110–11
Song-learning, 61
Strategy, 113–14
Utility, 30
Vacuum activities, 69–70
War of Attrition, 126–7